Who's Raising the Kids?

Also by Susan Linn

Consuming Kids: The Hostile Takeover of Childhood

*The Case for Make Believe:
Saving Play in a Commercialized World*

Who's Raising the Kids?

*Big Tech, Big Business,
and the Lives of Children*

Susan Linn

NEW YORK
LONDON

© 2022 by Susan Linn
All rights reserved.
No part of this book may be reproduced, in any form,
without written permission from the publisher.

Requests for permission to reproduce selections from this book should
be made through our website: https://thenewpress.com/contact.

Published in the United States by The New Press, New York, 2022
Distributed by Two Rivers Distribution

ISBN 978-1-62097-227-4 (hc)
ISBN 978-1-62097-228-1 (ebook)
CIP data is available

The New Press publishes books that promote and enrich public discussion and understanding of the issues vital to our democracy and to a more equitable world. These books are made possible by the enthusiasm of our readers; the support of a committed group of donors, large and small; the collaboration of our many partners in the independent media and the not-for-profit sector; booksellers, who often hand-sell New Press books; librarians; and above all by our authors.

www.thenewpress.com

Book design and composition by dix!

This book was set in Bulmer MT

Printed in the United States of America

10 9 8 7 6 5 4 3 2 1

FOR MARLEY AND IZZY, WITH LOVE

This society transforms its children into consumers making them want, want, want, want, in order to sell them and their parents not what the children need but what they have been made to want. What they want, not what they need. It commodifies and monetizes its children. It objectifies them. It dehumanizes them.

—Russell Banks

Contents

A Note to the Reader   *xiii*

    Introduction   1

1: What Children Need and Why Corporations Can't Provide It   11

2: Who Wins the Games Tech Plays?   33

3: And the Brand Plays On   57

4: Browse! Click! Buy! Repeat!   77

5: How Rewarding Are Rewards?   93

6: The Nagging Problem of Pester Power   109

7: Divisive Devices   125

8:	Bias for Sale	145
9:	Branded Learning	163
10:	Big Tech Goes to School	181
11:	Is That Hope?	197
12:	Resistance Parenting: Suggestions for Keeping Big Tech and Big Business at Bay	207
13:	Making a Difference for Everybody's Kids	229

Afterword	*241*
Acknowledgments	*243*
Appendix: Model Edtech Policy for School Districts	*245*
Suggested Reading, Viewing, and Listening	*249*
Resources	*253*
Notes	*261*
Index	*317*

A Note to the Reader

The facts are irrelevant. It doesn't matter one bit whether something is actually better or faster or more efficient. What matters is what the consumer believes.
—Seth Godin, *All Marketers Are Liars*

I was working away at this book in March 2020 as the COVID-19 virus spread, beginning to take its terrible toll around the world. Sheltering in place began. Confined to my house, I thought I would devote my unexpected, extended solitude to immersing myself in writing. Instead, I found myself distracted by the destruction at hand—mounting deaths, rampant unemployment, and beleaguered and inadequately provisioned health care workers. So rather than writing, I spent the first months of the pandemic engaging with young children in live video chats with my puppet, Audrey Duck—figuring I could at least provide a respite for kids abruptly deprived of their peers and for stressed parents working from home or suddenly unemployed.

The increasing evidence that Black and Brown people were disproportionately dying from COVID, the horror of George

Floyd's tortuous murder under the knee of a white policeman, and the wave of protests that followed only pulled me further away from writing. My concerns about immersing children in our hyper-commercialized culture seemed so beside the point.

But I was wrong. While out-of-control commercialism is not the root cause of the aforementioned problems, I came to realize that our marketing-saturated culture exacerbates these and other social ills. The mores and behaviors promoted by corporate marketing are no longer confined to commerce. They profoundly, and negatively, influence crucial realms of civil society, including government, family and community life, and schools and learning, as well as our relationship with ourselves, with others, and with nature.

In 2016 we elected a president previously known solely and simultaneously as a brand and its chief marketer. At the time, I thought this was the pinnacle of a market-driven society. But while Donald Trump may be the embodiment of a hyper-commercialized culture, he's also the result rather than the cause. A scary number of Americans still buy the lie that he won a bid for re-election that he empirically lost. And speaking of lies, we lost our best chance to thwart the coronavirus because pundits and politicians—aided by Facebook* and by other profit-hungry tech and media conglomerates—sold millions of people on blatant fabrications that denigrated real life-saving protections like vaccines and face masks.

Advertising targets emotions, not intellect, and is designed to forestall critical thinking. Iconic slogans like Nike's "Just Do It," Sprite's "Obey Your Thirst," and Pepsi's "Live for Now" glorify impulsivity. When I was in Korea several years ago, the slogan for Coke was "Stop Thinking."

I began to understand the United States' colossal and tragic failure to contain the coronavirus in its early days, potentially saving hundreds of thousands of lives, as a macabre affirmation of the

---

*Facebook changed its name to Meta in 2021. From now on I use "Meta" to refer to the company and "Facebook" to refer to the social network.

power of marketing. The Trump administration's handling of the virus certainly seemed to embrace the essence of Seth Godin's marketing maxim: facts are irrelevant. ("The virus is a hoax."[1] "It's something we have tremendous control over."[2]) What really matters is what people believe. The best way to sell something is to convince potential buyers that you're giving them what they need. People desperately wanted to believe that the government had the virus under control or that it would go away on its own—or better yet, that it didn't really exist.

Harms to children are collateral damage of the marketing techniques employed to spread disinformation about the coronavirus. While kids weren't the intended audience, they certainly suffered. At a minimum, children's lives were upended, their schooling interrupted, and their parents stressed. For many kids, the consequences were even more devastating—they lost parents and caregivers and descended into poverty. Some got sick themselves. Some even died.

I also began to understand that while the links between our tech-dominated, commercialized culture and systemic racism may not be obvious, they're real, they matter, and they affect children.[3] In addition, as authors Safiya Umoja Noble and Ruha Benjamin demonstrate so clearly, the algorithms powering profit-driven tech, including social networks and search engines, perpetuate racial and ethnic stereotypes.[4]

The racism sustained by commercialized algorithms harms not only those children who use social media and search engines but also millions of kids who don't. The stereotypes and misinformation that are a byproduct of algorithms that prioritize profit over justice profoundly affect how Black and Brown children are viewed and treated by society. In turn, how the larger society views various races and ethnicities affects how all kids view themselves and each other.

As I began to sort out the connections between excessive commercialism and the horrendous consequences of both a mismanaged public health crisis and systemic racism, my excitement about writing returned. Once again, it became clear to me that the problem with the tech-driven, omnipresent marketing that kids experience today isn't just that they're being sold stuff. It's that the values, conventions, and behaviors embraced and engendered by gargantuan, minimally regulated, for-profit conglomerates permeate all aspects of society, including the lives of children.

Instead of being beside the point, I've come to see that understanding and mitigating the impact of commercial culture on children, in particular the impact on their values, relationships, and learning, is vital to successfully moving through not only the crises I describe above but others we face now and in the future. What follows is my contribution to making that happen.

Who's Raising the Kids?

Introduction

Almost the entire children's media ecosystem—not all of it, but almost all of it—exists as a venue for advertising. Advertising and marketing are the foundation on which the whole media structure is built.
—Vicky Rideout, founder of VJR, specializing in research on children, media, and technology

I am giving a talk about kids, tech, and commercialized culture to parents, teachers, and administrators when I notice something strange. The people my age in the audience, whose kids are grown, are looking rather smug. In contrast, the younger generation of adults, those currently in the throes of raising families, are looking distinctly uncomfortable. So, I interrupt myself.

"I'm not here to make anybody feel guilty," I say. "In some ways, it's never been harder to be a parent, even for families with adequate resources. You're dealing with a culture dominated by multinational corporations spending billions of dollars and using seductive technologies to bypass parents and target children directly with messages designed—sometimes ingeniously—to capture their hearts and minds. And their primary purpose is not to help kids lead healthy lives or to promote positive values or even to make their lives better. It's to generate profit. So, if people my age tell you to 'just say no' or talk about how they used to just turn off the TV, be polite—but remember that they have no idea what raising children in a digitized, commercialized world is like." And I resolve to begin my talks this way from now on.

I'm beginning this book that way, too. My goal is not to make parents* feel guilty about failing to cope perfectly with the barrage of tech-driven commercialism engulfing their families. My goal is to help anyone who cares about children understand that the digitized, commercialized culture so many kids wake up to every day is toxic, that its impact extends beyond the wellbeing of individual children and families to the wellbeing of our larger society, that fixing it is a societal responsibility, and—importantly—that we *can* make things better.

As I describe in my 2004 book *Consuming Kids*, I've been working since the late 1990s to stop corporations from targeting children. The goals of marketers who target kids certainly haven't changed over the past twenty-odd years: lifetime brand loyalty, pathways to parents' paychecks, and the habit of consumerism. What *has* changed is that rapidly evolving digital technologies make it easier for companies to engulf children in advertising that is more pervasive, more invasive, more sophisticated, more manipulative, and more devious than ever. Meanwhile, evidence continues to mount that marketing to children is both remarkably effective and deeply harmful to their growth, development, and wellbeing.

My activism to prevent corporations from having unfettered access to children is rooted in both my personal life and my professional life. I am a mother, now a grandmother, and a psychologist.

---

*In these chapters, the word "parents" refers to guardians, and any adults responsible for raising children.

I've spent my entire adult life working for and on behalf of children. As a ventriloquist and children's entertainer, I was fortunate to work with Fred Rogers, first as a guest on his show *Mister Rogers' Neighborhood*. Later, I created video programs with his production company to help children talk about challenging issues like racism and homelessness. For many years I was a puppet therapist at Boston Children's Hospital and at the Children's AIDS Program out of Boston University Medical Center, helping kids cope with hospitalization, surgery, and life-threatening illness.

The deluge of commercialism aimed at children today has its roots in the 1980s when the deregulation of children's television made it legal to create programming for the sole purpose of selling toys combined with advances in digital technologies. By the 1990s I saw the impact of unrestricted commercialism on the young children with whom I worked and, closer to home, experienced it at my daughter's school. It was a four-year-old who introduced me to the pop singer Britney Spears. In fourth grade, my daughter's school devoted a whole semester of music class to learning Disney songs, the one body of music they were sold daily.

By the late 1990s I discovered that I was not alone in my belief that no one should manipulate children for profit. Grass-roots organizations began taking on commercialism in schools, sexualized toys, media, and clothing aimed at kids as young as preschoolers, and the negative impact of media violence on children. In 2000, recognizing that advertising directly to children is the link among all these problems, I and some colleagues formed the nonprofit advocacy organization Campaign for a Commercial-Free Childhood (CCFC), which is now called Fairplay. I served as its founding director until 2015, when I left the organization in the capable hands of Josh Golin, who is Fairplay's current executive director, and a dedicated board. I remain an adviser to Fairplay, and I continue to speak out about the myriad harms of marketing to children.

Over the years, Fairplay has successfully persuaded some of

the largest corporations in the world to change some of their most egregious marketing to children. To name a few of our triumphs: We convinced Disney to stop falsely marketing *Baby Einstein* videos as educational for babies.[1] We prevented Hasbro from releasing dolls based on the Pussy Cat Dolls, a burlesque troupe turned singing group, for six- to nine-year-olds.[2] We convinced the NFL to shut down Fantasy Rush, a fantasy football game targeted to kids.[3] And, working with advocacy groups, legislators, and the Federal Trade Commission, we stopped Google from collecting and monetizing children's personal information on YouTube Kids and forced the company to significantly limit the kinds of advertising allowed there.[4]

When The New Press approached me about updating *Consuming Kids*, I first thought that I would merely replace old examples of marketing with new ones and cover new research about the harms of marketing. But I quickly realized so much has changed in the interim that a mere revision wouldn't do. *Consuming Kids* was published years before smartphones and tablets drastically changed how entertainment and information are delivered. Now, whenever and wherever children happen to be, it's possible for them to be in front of a screen. I heard one marketer say recently, "The thing I love about the Lego apps is kids can take them everywhere! You can't take a bag of bricks everywhere." Apparently, he and others in what's now called the "kids tech industry" are thrilled that children can access screens at home, in the car, on the playground, at school, and everywhere else they go. What he didn't mention is the increasing evidence that excessive screen time is harmful to children's health and development.[5]

Meanwhile, digital technologies continue to evolve at breakneck speed, much faster than our understanding of the moral, ethical, physical, and social ramifications of their dominance in our lives—and the lives of kids. Today, software can accurately read emotion from facial expressions, seamlessly manipulate video images, and conjure increasingly real-seeming virtual worlds. It can also invest objects with the capacity to do our bidding and even cause us to love them.

Digital devices are marketed to parents as must-have child-rearing tools and marketed to kids as their sole opportunity for fun. In a rigorously antiregulatory political climate, tech and media companies target children with brilliant, sophisticated, ubiquitous, and often obfuscated marketing that is seamlessly integrated into digital content and programming that is created purposely to be addictive. In the tech world such manipulation is called "persuasive design." Rooted in behavioral psychology, persuasive design is the science of programming computers to alter human behavior.

The astounding science fiction–like capabilities of computers are themselves only part of the problem. In the United States, the rise of powerful digital technologies parallels the decline of government regulations designed to set limits on excessive and irresponsible corporate behavior. In combination, these two phenomena drive a societal embrace of commercialism that powerfully affects children, both directly and indirectly.

One major direct influence, for instance, is the huge popularity of social networking sites, which transform us and our children simultaneously into our own brands and into marketers for corporate brands. The selfies we post on platforms like Facebook, Snapchat, and Instagram can easily be digitally altered to erase physical flaws, real and imagined, just like marketers enhance photos and videos of the objects they sell. What we and our kids choose to post or not post online shapes our personal "brand."

As I write, one of the most popular online activities around the world for young children is watching other kids open boxes of toys on YouTube, Google's popular social media platform. Many of the seemingly spontaneous demonstrations in these "unboxing videos" are paid for by toy companies. At a recent marketing conference, I heard toy company executives extol the power of working with the kids they recruit, whom they call "influencers." Not only do these videos market products to children they also shape aspirations. A few weeks ago, I spoke to a father who told me that his six-year-old son desperately wants to be a YouTube star. A grandmother recently shared her anxiety about her twelve-year-old granddaughter, who was posting videos of herself on YouTube. Here's what she told me: "My granddaughter is addicted to YouTube. She sees other little girls in the spotlight so she wants to make her own videos. And one day I found her making one when she wasn't fully dressed." A pre-teen, her granddaughter is tech savvy enough to bypass age restrictions set by social networks. But "tech-savvy" does not mean that her judgment is developed enough to grasp the potential consequences of her online behavior.

And she's certainly not alone. While I chatted with a fellow guest at a wedding recently, she shared this story about her six-year-old niece, whom I'll call "Olivia." She said, "Olivia was playing in her room with her ten-year-old neighbor. When Olivia's mom checked on them, she discovered that the neighbor was about to post a video on YouTube of Olivia in her underwear. What if her mother hadn't walked in at that time?" These children were prevented from making a potentially devastating mistake. But not all kids are that lucky. In 2020, 15 percent of nine- and ten-year-old children reported having shared a nude photo of themselves online.[6]

It's not unusual for children to be occasionally cruel, to be titillated by nudity, and to be unable to foresee the consequences of their actions. Today, given the reach of social media, what used to be children's relatively private meanness, explorations, or mistakes may now be made public. These childhood transgressions might be witnessed not only by friends, families, and communities (which is shaming enough) but by millions of strangers—including predators.

It's commonly accepted that children's health, behavior, and values are affected by their encounters with direct influences—their experiences with family, school, and community, and in recent years, with media. But children are also affected by indirect societal forces—the economic, cultural, religious, and political systems in which they live. Today, the institutions serving our spiritual, academic, civic, and social lives actively incorporate the language, values, and techniques of marketing.

Our children and grandchildren are growing up in a culture that blurs the boundaries between public and private, civic and commercial, philanthropic and profit-making. Churches, synagogues, and mosques are urged to hone their "brand." Grocery chains position themselves as public benefactors by asking us to pay extra to support their charitable causes at check-out counters. Charter schools can be run by corporations, and public schools take corporate donations in exchange for branding children's learning. Tech companies can provide schools with online access in exchange for collecting personal data on students. Even our experiences with nature are becoming branded. Advertising is now allowed in our national parks.[7]

In *Consuming Kids*, I wrote primarily about the links between commercialism and a host of children's public health problems, including childhood obesity, eating disorders, precocious sexuality, youth violence, family stress, and the erosion of children's creative play. In this book, I explore the impact of tech-enabled marketing on what anthropologists call "deep culture." I'm using that term to define the aggregate of underlying, and sometimes unconscious, convictions, values, and attitudes that influence how we conceptualize ourselves and others, that motivate us to act, that prompt the choices we make and don't make, and that reflect what matters most to us.

When thinking about the effect of commercialism on children, it's crucial to remember that advertising sells values and attitudes as well as products. Children in a hyper-commercialized culture are continually sold the belief that the things we buy will bring us happiness. Research has long shown us that the things we buy *don't* make us happy in any kind of sustained way, which can leave people who believe they do in a constant state of disappointment and dissatisfaction.[8] This belief surrounding consumerism works well for the companies making these things, since believers buying an object who find that they don't stay happy often blame the object and search for a bigger, better, more exciting thing to buy. In fact, research suggests that children with more materialistic values are *less* happy than their less acquisitive peers.[9]

Other harmful messages are also embedded in a culture that promotes consumption as a path to happiness. These include the celebration of extrinsic—rather than intrinsic—motivations and rewards, as well as instant gratification, self-importance, impulse buying, and uncritical brand loyalty. As I'll describe later, these tenets of commercial culture don't just undermine children's wellbeing, they also threaten democracy and the health of the planet.

The chapters that follow expand the case against marketing to children, highlight the need for regulations to stop it, and advocate for massive parent and public education to help us keep commercialism at bay. I begin with a look at what we know children need to thrive and the habits and attributes they need to lead healthy, meaningful lives. I explore digital technologies and how tech industry business models undermine children's wellbeing and healthy development. I share what I've learned from attending marketing and tech conferences, including how people in the business of marketing to kids talk about children and families. And I describe the techniques companies use to hook children and their families on their products. I look at the gargantuan edtech industry and its takeover of children's time in school. I explore the ways that companies like Google and Amazon exploit advances in artificial intelligence to insert products between parents and children that disrupt the development of normal attachment and encourage kids to rely on tech and not on parents or friends or themselves for soothing, amusement, and learning about the world.

In addition to laying out the problems, I include suggestions about what we can do at home, in schools, in communities, and at a policy level to provide children with the screen-free, commercial-free time that is so crucial to their wellbeing. I also offer resources for parents, educators, health professionals, and advocates.

One truism about what advertisers call "the kids' market" is that it changes rapidly—especially in this digital age. So, it's possible that when you read this, the specific platforms, applications, games, and toys I describe may be old news, or even defunct. But these products and the business models that propel them remain relevant because they continue to be emblematic of how and why children's wellbeing is threatened when kids are left unprotected in the marketplace.

Finally, most of my experience working with and on behalf of children has been with neurotypical kids coping with difficult physical and psychosocial challenges. So when I talk about children in the following chapters, I'm referring primarily to neurotypical kids. I absolutely recognize the need for addressing the impact of tech and commercialism on neurodivergent children, but it's beyond the scope of this book.

What's happening to children in our digitized, commercialized world is deeply distressing. The fact that it's made possible by the biggest corporations in the world is certainly daunting. But the possibilities for effecting change are by no means nonexistent. Profound social change takes time. What's happening among advocates and activists working to stop the corporate takeover of childhood and to promote environments that encourage healthy development is both hopeful and exciting. And at least some of the hope and excitement is coming from within the tech industry. There are tech executives and engineers taking on their own industry, calling out abuses of privacy and identifying the need to protect children from the seduction of persuasive design. The fake news epidemic made possible by Facebook's lackadaisical self-regulation lays bare many of the dangers of reliance on social media. Hacks of products from children's tech brands such as VTech and CloudPets and countless other security breaches have made millions of people more aware of and concerned about privacy violations.[10]

As more research emerges about the potential harms of excessive screen time, educators, health professionals, parents, and legislators are starting to address the need to set limits on prevailing tech industry business practices that target kids to ensure that children have time for what they really need—hands-on exploration of the world around them, face-to-face conversations with family and friends, active and creative play, both indoors and out.

It's my hope that this book will help readers recognize the need to stand up to the corporate interests that highjack children's lives—and help ensure that kids have the childhood they need to thrive.

1

What Children Need and Why Corporations Can't Provide It

The secret sauce is not fancy toys and computers and electronics. Things that allow their imagination to run wild like play and relaxation. That's what builds a really good brain.
—Kathy Hirsh-Pasek, co-author of *Becoming Brilliant: What Science Tells Us About Raising Successful Children*

I am watching a twenty-five-second video that takes my breath away. Arielle, my cousin Ellen's fourteen-month-old granddaughter, sits on a rug with an old baby doll, a stuffed bear, and a couple of books. Notable for their absence of buttons to push or screens to swipe, these objects neither talk, sing, chirp, beep, move, nor play music. They merely lie there, waiting for someone to do something with them. Arielle explores the baby doll while making the only sound in the room, a combination of crooning and babbling. She chews meditatively on the doll's arm for a bit, drops it in order to hold up its foot with one hand and run a finger over its toes with the other. She reaches up to feel her ear and is momentarily stymied. Something doesn't compute.

Dropping the doll's leg, Arielle's hand wanders up its torso until it encounters a tiny ear. She bends over, using one finger to trace its contours. Reaching up, she first feels one of her own ears and then both ears simultaneously. She alternates between tracing the doll's ear and her own a few more times until, satisfied, she turns her attention elsewhere.

I am witnessing a paradoxically astonishing and completely ordinary feat of human learning—at least for neurotypical kids in safe, loving environments. Something piques Arielle's curiosity—is her body like her doll's body? With no outside prodding or direction, she initiates the process of satisfying that curiosity (feeling the doll's body and her own). When her initial attempt fails (the doll's toes and Arielle's ear are not similar), she persists in her quest for an answer by trying out another possibility (finding the doll's ear and comparing it with her own).

Arielle's moment of discovery is emblematic of so much that children need to flourish. She is playing in a safe space, in the presence of someone who loves her, with access to objects that invite her to use them as she wishes instead of dictating how they should be played with. As a result, she gets to experience curiosity and then the satisfaction of figuring out, on her own, the answer she is looking for. She gets to experience a very manageable failure and persevere until she completes the task she has set out for herself.

What isn't overtly visible in our twenty-five-second glimpse into Arielle's life, but is crucial for her development, is that all her basic needs are met. The underpinnings of healthy growth and development lie in the adequate satisfaction of children's need for food, love, and both physical and psychological safety. Children's lives are stunted if they are chronically hungry, frightened, hurt, or unloved. From infancy, kids need food, safety, and consistent connections with at least one adult who loves them and responds not only to their physical needs but also to their social and emotional needs. Our first obligation to children is to make sure these basic needs are met. But our obligations don't stop there.

Thanks to advances in neuroscience and recent research in developmental psychology, we now know a great deal about what young children need to thrive and why the first years of life are so important. My colleagues Joan Almon and Diane Levin and I put it this way in *Facing the Screen Dilemma:*[1] "Babies begin life with brains comprised of huge numbers of neurons, some of which are connected to each other, and many of which are not. As children grow and develop, everything they experience affects which neurons get connected to other neurons. Repeated experiences strengthen those connections, shaping children's behavior, habits, values, and responses to future experiences. The experiences young children don't have also influence brain development. Neurons that aren't used—or synaptic connections that aren't repeated—are pruned away, while remaining connections are strengthened. This means that how young children spend their time can have important lifelong ramifications. For better or worse, repeated behaviors can become biologically compelled habits."[2]

It's commonly accepted that chronic family stress and disruptions, wars, poverty, inadequate schools, racism, and unsafe neighborhoods can deprive children of experiences that nurture healthy skills and attributes and subject them instead to experiences that encourage harmful habits and behaviors.[3] Here's what's rarely acknowledged: ongoing immersion in a commercialized, digitized culture can also deprive young children of experiences that nurture the skills and attributes they need to flourish—like creativity, curiosity, and agency. While no decent person would ever wish chronic family tensions and disruptions or other known stressors on kids, too often well-intentioned adults welcome commercialism into children's lives from birth and even fork over lots of hard-earned cash to ensure their immersion in it.

Brand-licensed toys, clothing, and accessories account for about 25 percent of children's products.[4] That means they are de facto ads for something else. Brand-licensed toys feature, for instance, beloved media characters like Elmo or Spiderman and promote the movies, videos, TV shows, and apps showcasing these characters, as well as any similarly branded food, clothing, or accessories. And the movies, videos, TV shows, and apps showcasing characters promote the products featuring their images.

What's equally concerning is that much of what kids play with today are digital devices and electronic toys. In one survey, when asked what their kids had at home to play with, most of the parents (77 percent) listed products like game consoles, smart phones, and tablets. Electronic toys were identified by 63 percent of parents.[5]

To explain why I worry that commercially driven electronic devices and toys dominate the leisure time of young children today, I am compelled to turn, as I often do, to D.W. Winnicott, the mid-twentieth-century British pediatrician and psychoanalyst whose work continues to influence my understanding of what children need and don't need. Winnicott is perhaps best known for coining the term *good enough mother*, which frees parents from the burden of believing they must be perfect. What matters most to this discussion, however, are his writings about play as crucial to developing a healthy sense of self. Play is a conduit for true self-expression. It facilitates the ability to initiate action rather than merely react to stimuli, to wrestle with life to make it meaningful, and to envision new solutions to old problems. Winnicott believed, as do I, that the ability to play is a central component of a life well lived.

Play seems like a simple concept, but it is notoriously hard to define. Play in the Winnicottian sense is a paradox. It is active, self-initiated, self-driven, creative, and voluntary. It is focused on process, not results. Yet great art, literature, music, scientific discoveries, and other human accomplishments emanate from play with ideas, materials, and concepts. As Winnicott says, "We play in the service of the dream." Play has to be fun, yet it can involve the expression of serious, profound, and even frightening thoughts and experiences. We can play alone or with others. All play has rules and structure, even if they are unstated. And since when we play with other people everyone has to agree on the rules, play necessitates cooperation and interpersonal communication. Competitive games can be play, but not when winning or losing becomes more important than the process of playing.

As I have written elsewhere, Winnicott believed that play flourishes from infancy in the context of environments that are simultaneously secure enough to be safe and relaxed enough to enable spontaneous expression. The different ways we can hold and interact with babies provide a good metaphor for the impact of the environment on play. A baby held too loosely doesn't feel safe to generate movement and must remain perfectly still for fear of falling; a baby held too tightly has no space to generate movement and never has the experience of originating action. However, when babies are simultaneously held securely enough to be safe and loosely enough to have some freedom of movement, they experience what it's like to generate action—to choose to make a gesture. Winnicott calls environments that are safe but allow for freedom of expression "holding environments." It's a great term. A caretaker's arms can be a holding environment. So can a relationship. So can a family. So can a classroom.

Think about the different possible ways to respond to a baby. When a baby makes a gesture that seems purposeful or that suggests independent personhood, we often respond—without even thinking about it—with a coo or a smile or a laugh. In that interchange are the seeds of two important developmental changes essential to creativity.

By originating an action that evokes a distinct reaction from her environment, the baby begins to learn to differentiate herself from her parent and to establish a sense of herself as a separate being. The knowledge that we are separate from our environment and from the people around us is an essential foundation for healthy growth and development. If, as babies, our early actions evoke loving responses from the important adults in our lives, we experience our burgeoning self as able to make good things happen in the world. We experience creativity.

An inadequate holding environment may be one where babies are neglected or where their safety is compromised. Suppose the baby makes a gesture and no one responds? Suppose she generates a gesture that elicits anger instead of support?[6]

The environment that is most relevant to our discussion, however, is what Winnicott describes as "impinging." He was talking about when people are so busy eliciting responses from a baby ("Do this!" "Smile!" "Do that!") that she has no space to even try to generate an action on her own. How does she ever learn who she really is? Creativity, or constructive spontaneity, in contrast to the constant compliance or reactivity demanded by an impinging environment, is at the core of his conception of the difference between mental health and illness. A child growing up in an environment that is either unsafe as the result of neglect or too demanding as the result of incessant stimulation and demands to respond may develop a reactive or "false" self instead of the "true" or creative self that flourishes in a supportive holding environment.

As babies develop, their environment expands far beyond the safety of nestling in our arms. But their need for "holding" in order to play spontaneously and creatively continues both in their physical environments and in their relationships. Children need to be both physically and emotionally safe. They need room to explore and experiment within boundaries that protect them from harm but do not constrict. They need physical space in the form of safe play areas, and they need adult relationships that simultaneously nurture freedom of expression and delineate clear boundaries that keep them from being either hurt or hurtful. Children also need silence and time on their own in order to experience the difference between reacting to outside stimulation and generating their own ideas in play.[7]

For children growing up today, unregulated commercialism—motivated primarily by hunger for profits and delivered by ubiquitous, seductive technologies—constitutes an impinging environment on steroids. With help from corporate-supported psychologists, artists, engineers, researchers, tech moguls, and more, out-of-control commercialism can overwhelm kids and squelch the true self that Winnicott equates with health.

Watching Arielle make her momentous discovery and thinking about Winnicott's concept of an impinging environment, I was struck by the contrast between her experience and the experience of infants and toddlers I had met not long before at a local day care center. The children were on the floor, surrounded by the kinds of bestselling toys marketers today promote as "interactive," in that they move and emit sounds at the push of a button or the flip of a switch. The room echoed so much with sounds of tinkling music, whistles, beeps, and robotic voices reciting the ABCs or nursery rhymes that it was hard to think. It took me a moment to realize that all the noise and activity emanated from the toys. In contrast, each child in the room sat silent and passive, as if stunned by the electronically enabled commotion.

I've noticed a similar kind of response in babies and toddlers using phones and tablets. Have you ever looked at YouTube videos of babies and toddlers using iPads on their own? Most of these tiny children are so focused on the screen that they seem oblivious to whatever else is going on around them. They repeatedly, rapidly, and randomly tap, swipe, and open and close their little fists with abandon, completely absorbed in the screen's constantly changing images. What appears to matter most to babies is not making sense of what's on the screen but exercising their power to alter the images as quickly as possible.

"What you say is true, in my experience," the father of two young children writes to me, "but what makes it so different from the kind of off-screen play you like so much? In both cases babies are exploring the world and experimenting with what they encounter—they're just doing it on a screen." It's a question I get a lot. One important difference is that research on babies, toddlers, and tech suggests compelling reasons to postpone children's exposure to screen-based entertainment until after infancy and toddlerhood.* In contrast, scads of evidence show the benefits of hands-on, creative, child-driven, and tech-free play like Arielle's exploration of her doll.[8]

As for me, I worry when I encounter babies and toddlers immersed in screens. There's no question that screen-based moving images can immediately and intensely attract their attention. In fact, they seem so engaged that they rarely look away from the screen. But "engagement" in an experience doesn't necessarily mean it's either useful or meaningful. For babies mesmerized by phones or tablets the "world" they're concentrating on is confined to a flat rectangle smaller than a square foot. It's a world without taste, smell, textures, and—most important—other people. It's also a world mediated by decisions made by someone else about what they should look at. Contrast that with the baby I encountered the other day at my local farmer's market. She alternated her attention between the sights and sounds around her and an absorbing interest in her feet. Every so often she would reach down and grab on to her foot, bending first one leg and then the other. It reminded me of a time when my daughter was nine months old, and I was pushing her in a stroller. Suddenly, and I have no idea why, she decided to bend over and, for a block or so, watch the world going by beneath her stroller. Like Arielle's unprompted fascination with her doll's ears, these small, seemingly trivial, acts of volition—of exercising and satisfying innate curiosity—are marvelous examples of burgeoning independence and personhood.

It seems to me that from the beginning of their lives, we want babies to be endlessly curious about what's happening around them—not just what they see looking a few inches in front of them. And, as with the electronic toys I describe above, apps purportedly designed for babies and toddlers bombard them not just with

---

*I describe research on babies, toddlers, and tech in chapter 2.

moving images but also with sound effects and spoken words. The apps do so much there seems little motivation for a baby to experiment with vocalizing. I once downloaded a YouTube video of a toddler playing with an iPad in which a voice on the app enthusiastically exhorts users to "Use your words!" She doesn't.

Any toy, game, or app that speaks or emits sounds deprives children of the pleasure of making up what they say and the fun of vocalizing sound effects and silly noises. The hot holiday toys in 2021 included "My Little Pony: A New Generation Sing 'N Skate Sunny Starscout Interactive 9-Inch Remote Control Toy with 50 Reactions, and Lights" ($52.99 at Target).[9] Then there's the "Magic Mixies Magical Misting Cauldron with Interactive 8-Inch Blue Plush Toy and 50+ Sounds and Reactions, Multicolor" ($69.99 on Amazon). And, from the popular YouTube show for preschoolers, "CoComelon Official Deluxe Interactive JJ Doll with Sounds" ($59.88 at Walmart).[10]

Electronic wizardry makes for great advertising campaigns. The toys look fun. But they are created with a kind of planned obsolescence. They are not typically designed to engage children for years, or even weeks. They are designed to sell. If interest wanes, so much the better—another version will soon be on the market. These toys are great for profits. But great for children? Not so much.

The more a toy or app drives the form and content of children's play and the more the characters or the toys kids play with are linked to popular media properties and franchises, the less children get to exercise curiosity, initiative, creativity, flexible problem-solving, and imagination. They are more likely to react rather than generate and less likely to set their own challenges and figure out how to meet them.

Children are born with innate capacities for skills and attributes that enable them to embrace prosocial values, to engage in meaningful relationships, and to immerse themselves in active learning. I believe adults are obligated to ensure "holding" environments and relationships that liberate these capacities and nurture their development. I also believe we have an obligation to protect children from environments and relationships that prevent or delay these capacities from developing. I believe in this universal obligation for the sake of each child as an individual and for the benefit of society and the planet. Based on current research, theories of whole-child development, and my own experience with kids, what follows are the ones on my (admittedly) abbreviated list that are particularly relevant to this discussion.

Qualities like kindness, empathy, generosity, and compassion contribute to satisfying, meaningful relationships with our fellow humans and ourselves. These are also important components of democratic values like justice, equity, and fairness. Kids develop these attributes first by experiencing them in relationships with primary caregivers who value and try to embody them. In other words, children learn how to treat other people based on both how they are treated and how they see the people around them are treated.

One reason the content of children's entertainment is so important is that its characters serve as models for how children behave. This was made embarrassingly clear to me early in my performing career. I was performing for first-graders, and I wanted to do a piece about how name-calling hurts feelings. I decided to make my puppet Audrey Duck call her counterpart, Cat-a-lion, "stupid." I planned to spend most of the time addressing the consequences, including hurt feelings (Cat-a-lion) and shame and guilt (Audrey). The piece began splendidly. When Audrey called Cat-a-lion "stupid," the kids were enthralled. But suddenly I was interrupted. Someone in the audience began to mimic Audrey and call Cat-a-lion "Stupid!" To make matters worse, one after another, the other kids joined in. Suddenly the whole audience was yelling "Stupid!" at Cat-a-lion. Needless to say, it wasn't the response I expected. With the best of intentions, I had just, by example, given kids permission to call other people "stupid!" I never did *that* again.

Curiosity, initiative, persistence, creativity—all are foundational for learning, for constructive problem-solving, and for the ability to follow tasks and challenges through to completion. As children's brains mature and develop, so does their facility with two important skills. One is self-regulation, which includes the ability to delay gratification and to exercise control over impulses and emotions—or, as the adage goes, to think before we speak (or act). The other is critical thinking, which embodies the will and ability to differentiate fact from fiction.

Self-regulation and critical thinking each broaden and deepen our life experience. They are essential for being part of a community, for civil society, and for accomplishments in the realms of art, sciences, humanities, and for devising solutions to thorny political and social problems. Both skills are central to a peaceful, functioning democratic populace.

A latecomer to my list is the capacity to feel awe, which in childhood first manifests as wonder. It's a complicated emotion because it can encompass both joy and terror. Great beauty and great devastation can each engender awe. Researchers who study awe in adults find that people describe the feeling as a sense of vastness, engendered by encountering something that we can't quite assimilate and that therefore expands our sense of the world. People describe being awestruck as feeling simultaneously small and a part of something larger than themselves—they feel both humbled and connected.[11]

It's commonly accepted that wonder comes naturally and more frequently to young children than to adults. That makes sense. Young children constantly encounter brand-new experiences that expand their understanding of the world. Kids don't just feel small in a vast world, they actually *are* small and—if they are fortunate—their smallness isn't terrifying because they feel secure in their connection to the people who love them.

That children abound with wonder is one reason to delight in their company. As I wrote in my book *The Case for Make Believe*, children view with fresh eyes that which we take for granted. Even tasks that seem mundane to us can be wonder-filled for children. Undressing, for instance, except as a prelude to making love, is certainly among the most unremarkable of experiences—an activity necessary to transition to something else: showering, going to sleep, dressing for work or play. When Cassidy, at thirty months, figured out how to remove every scrap of her clothing, she burst into the living room totally naked. Standing before her bemused parents, she flung her arms out in triumph and proclaimed rapturously, "I escaped my clothes!"

The ability to remove her clothes is a thrill and a source of wonder for Cassidy. She's amazed at her competence and revels in her newly won freedom. Until she learned how to rip apart Velcro and snaps, or maybe even unzip or unbutton garments, she literally was trapped in her clothes. Soon she will delight in mastering the even more complex task of putting clothes on, and she will seek dominion over what she wears. "I dressed all by myself," a four-year-old announces with great pride. "Even the buttons!" It's not a given that we leave wonder behind as we outgrow childhood. In fact, wonder is an essential component of creativity. The story of Sir Isaac Newton getting bopped on the head with an apple and coming up with the theory of gravity seems to be apocryphal, but it's a good metaphor for the necessity of wonder in scientific discovery. Up to that point, we were all tooling along in life taking for granted the fact that objects fall. Discovering new ways of understanding the world involves a first step of recognizing the extraordinary in what others see as ordinary, or don't see at all. Albert Einstein's sense of wonder, for instance, that a compass always points north led him, as an adult, to explore invisible forces such as magnetic fields.[12] Art also depends on wonder. One quality that differentiates great visual artists from excellent draftsmen is the unique way artists see their environment—the quality of light illuminating a leaf, the pattern left by waves on sand, the droop of skin on an aging face. They notice things about the world that the rest of us miss.

Awe, or wonder, is also essential to spirituality. Abraham

Joshua Heschel, a great twentieth-century rabbi, activist, and philosopher, coined the term "radical amazement," which begins with a sense of awe that the world around us exists and that we are a part of it—to wonder at the miracle of our own breathing; at volition; at the majesty of the natural world, and at our place in it, including the recognition of that which we can and cannot control.[13] But Heschel also asks us to wonder at our capacity for wonder. It's certainly amazing that the dogwood trees outside my window blossom year after year, or that a blossoming clematis vine climbing up my porch can sprout from a single seed. But it's also amazing that I am alive and able to experience that amazement. In those moments—amid the strain and distractions of daily life—when we are able to grasp the wonder of our wonder, we experience life with heightened consciousness.[14]

These experiences of awe both deepen and broaden not only our lives but also the lives of children. For this reason alone, I believe that enabling wonder to flourish in children is important. It's also true that society at large benefits when people have opportunities to experience awe. Studies suggest that after an awe-inspiring experience, people tend to be more kind, more generous, and less materialistic. They care less about money and more about the environment.[15]

Psychologist Dacher Keltner, the founding director of the Greater Good Science Center and a pioneer in the study of awe, explains the links between awe and prosocial behaviors this way: "The experience of awe is about finding your place in the larger scheme of things. It is about quieting the press of self-interest. It is about folding into social collectives. It is about feeling reverential toward participating in some expansive process that unites us all and that ennobles our life's endeavors."[16]

As they are for many people, some of my own experiences of awe are connected with nature. I felt awe when I saw a total eclipse of the sun, when I had an up close and personal display of two blue-footed boobies intent on their mating dance, or the time by the ocean when I happened on a pair of dolphins breaching in perfect synchrony, or when I sat in a canoe and watched a mother bear take excellent care of her twins as they cavorted on the shore.

Our inspirations for awe don't have to be grand. They can be found in our daily experience. I am often awed by the exquisite humanity of very young children—by their volition and their universal drive to learn, to connect, and to make sense of the world. I got goosebumps when I first watched Arielle's quest to understand her body in the context of her doll's body. The first time my baby daughter pointed at me and said "Mama" filled me with awe. So did the time before she could really talk when she managed to communicate quite clearly that she'd left a cookie in our den and couldn't get it because the door was closed. Encountering the universality of human development leaves me awestruck and, as the research suggests, both humbled and deeply connected to humanity at large.

Rachel Carson, author of *Silent Spring*, the 1952 book that is often credited with launching the environmental movement, cared deeply about nurturing wonder in childhood, especially in connection with nature. Her last book, *The Sense of Wonder*, written to introduce children to the wonders of nature, beautifully describes her explorations in the woods with her three-year-old nephew. She believed that for children to sustain their inborn sense of wonder, they need at least one adult who can share it and rediscover the "joy, excitement and mystery of the world we live in."[17] The crucial word here is *share*. She doesn't tell us to instruct kids when we're out in nature with them, or to impinge or impose on a child's experience, but rather to allow it to unfold in quiet.

So much of children's commercial culture is rife with noise. I've never been able to quiet the kinds of electronic toys I watched overwhelm those children in the infant and toddler room. And while most of the children's apps I've played with can be muted, their default usually is to make sound. Today, children's opportunities for silence—to experience wonder but also to play, dream, and explore—are rare.

It was a long-ago conversation with Fred Rogers that first got me thinking about the importance of silence in children's lives. Silence was so important to him that he once used an egg timer to tick off a whole minute of it on his television show.[18] And after listening to cellist Yo-Yo Ma play his cello, Fred commented, "After you've heard someone play beautiful music, sometimes you just like to have a quiet time to remember it. Let's just sit and think about what we've heard."[19]

Children need access to quiet times and spaces. Not only is incessant noise a threat to physical and psychological health,[20] it is an internationally recognized form of torture.[21] Silence enables children to find their own voice, and I mean that both literally and figuratively. One ingredient of Arielle's exploration of her doll's body is that the only sounds in the room were of her own making. The vocalizations she produced are important precursors to language. And like the exploration itself, in choosing to use her voice just for the sheer pleasure of it, she experienced autonomy and the rudiments of verbal self-expression. That she was playing in a room in the presence of an adult who allowed her periods of silence gave her the opportunity to listen to her own thoughts and to transform them into action.

Arielle profited enormously from generating her own ideas about what to do with an unbranded old baby doll. Perhaps the overarching problem of commercialized play and playthings is that they deprive kids of opportunities to exercise their own imaginations. As the saying goes, "The best toys are 90 percent child and only 10 percent toy."* When it comes to toys, "best" needs to be differentiated from "bestselling." The bestselling toys, often those that are most advertised to kids, are often digitally enhanced or linked to popular media characters, or both.†

---

* I first heard this from the late Joan Almon, who directed the Alliance for Childhood for many years.
† There are notable exceptions. In 2020, one was MGA Entertainment's LOL doll brand, which is problematic in its own way and which I discuss in chapter 4.

From the standpoint of toy industry profits, Arielle's creative play and explorations were a bust. That's why allowing corporations unfettered access to kids threatens healthy development. After all, what corporations and children each need to thrive is different and, more important, often mutually exclusive. Corporations have an unending and insatiable need to generate profit for shareholders, which almost always trumps all other considerations. Take the food industry for instance. Junk foods, high in sugar, salt, and fat, are major moneymakers. They are also known to contribute to childhood obesity, which continues to be a major public health problem. Yet, despite years of evidence-based consumer activism, corporate manufacturers and purveyors of junk food continue to market these products to children.[22]

In the world of entertainment, sex and violence are big sellers. That's why companies that make violent video games, films, and TV shows still hawk their wares to children despite decades of research demonstrating that exposure to violent media is a risk factor for children's aggression, lack of empathy for victims, and positive attitudes toward violence as an acceptable solution to conflict.[23] Sexualized toys, media, and clothing are also big sellers despite evidence that they can be harmful for young girls.[24]

In contrast, neither the environments children need to flourish nor the skills and attributes they need to acquire are particularly profitable, at least in a financial sense. And the competing needs of children and corporations are especially evident when it comes to play. I and many others have written extensively about the critical importance of children's creative, hands-on, and active play. It's the foundation of learning, creativity, constructive problem-solving, and how they wrestle with life to make it meaningful.

For corporate marketers, and the companies whose products they sell, dominating children's playtime is the most direct and powerful route to capturing children's hearts and minds. Convincing kids and their parents that having fun depends on purchasing an ever-changing cadre of the latest toys is a gold mine for the companies that manufacture those toys, for media companies whose characters are featured on the toys, and for tech companies looking to hook kids on screens.

When I give talks on play, I often bring out three puppets, one at a time, and ask people to tell me what each is, its name, what it says.[25] The first is a puppet whose species, gender, and personality are appealing but purposely indeterminate. The second is what looks to me like a horse. The third is the beloved *Sesame Street* character Cookie Monster. The audience invariably can't agree on anything about the first puppet—some think it's a worm, some a creature, others call it a friend. Some see it as male and some see it as female, and so on. The possibilities are endless, because ideas about who and what that puppet is are generated by its beholder. These qualities don't emanate from the puppet itself. When I show the next puppet, almost everyone agrees that it's a horse, mule, or donkey. So already their creative choices are being limited. Yet there are still lots of options for its name, gender, voice (besides "neigh" or "hee-haw"), and so on.

Finally, I bring out Cookie Monster, and an instant "aww" fills the room. Everyone agrees that it's Cookie Monster, and when I ask what Cookie Monster says, they inevitably respond with great gusto, "Me want cookie!" Clearly this is a beloved, universally recognized, and utterly familiar character in their lives. Ironically, that they are beloved and familiar renders toys like the Cookie Monster puppet, which replicate ubiquitous media characters, a barrier rather than a pathway to creative play. Children play less creatively with media-linked toys.[26] If Cookie Monster were an electronic version that moved, talked, and played music on its own, it would be an even greater barrier to creative play.[27]

Have you ever tried to convince children that a character they know, love, and see repeatedly on screens can deviate from its predefined parameters? It's difficult. I remember sitting on the floor of a playroom at Boston Children's Hospital playing with a five-year-old named Annie. She found a set of plastic dinosaurs, took one for herself and handed me another one, announcing, "These have to fight." I wasn't thrilled with the idea of making my dinosaur fight hers, so my dinosaur began to talk. "No!" Annie insisted. "They can't talk." I was puzzled. "Why can't they talk?" I wondered. "It's like the movie," she explained impatiently. "You know! They fight and they can't talk!"

The problem of media-linked toys goes beyond the way they inhibit children's creativity while they play with the toys. What's more worrisome for children—and beneficial to corporate profits—is that a steady diet of immersion in programs and a dependence on the toys they sell convinces kids that they need to have seen a program to know what to do with a toy and that the only toys worth having are the ones linked to media programs.

When I met four-year-old Sophia, she wore a purple dress branded with the TV series *My Little Pony* on its front and she was carrying a *My Little Pony* lunch box. As we started to play, she brought out three stuffed toys. One was Vampirina, who stars in her own animated show on Disney Junior. Another was a fish named Dory, who is a supporting character in Disney's 2003 hit *Finding Nemo* and who catapulted to stardom in its 2016 sequel *Finding Dory*. The third was a plush Hello Kitty, a character created by Japanese merchandising company Sanrio in 1974 to adorn children's items and that exploded into an international sensation with her own TV shows and amusement parks.[28]

I asked Sophia where each of them lived. She explained that Dory lived in the ocean and that Vampirina was born in Transylvania and moved to Pennsylvania. The details she offered about the two characters were prepackaged by the brand. When I asked her about Hello Kitty, however, she was stumped. "I don't know," she said. "I haven't seen the program." Since she'd never seen a Hello Kitty program, she literally couldn't imagine where the character might live, and the possibility that she could invent a home, neighborhood, or land for Hello Kitty didn't occur to her.

I don't think that Annie's and Sophia's dependence on commercial culture to tell them how and what to pretend is so unusual these days. When I talk to preschool teachers, I often hear concerns that children's creative play is declining. And what research exists seems to bear this out. Creativity in the United States rose steadily until the 1990s, and then it began to decline—particularly among young children.[29] Of course it's possible that all sorts of factors contributed to an overall decline in children's creativity. Children's lives are more structured than they used to be, organized sports have replaced free play for many children, and art, music, and recess have been cut from many public schools. It also seems likely that children's increased immersion in commercialism is a factor in their decreased creativity.

The decline began in 1990, ten years after the Federal Trade Commission lost most of its power to regulate marketing to children and six years after the Federal Communications Commission, during Ronald Reagan's presidency, deregulated children's television,[30] legalizing creating programs for the sole purpose of selling toys.* As a result, whether a program could move merchandise became more important to its survival than the program's content or its messages. By 1985, for the first time, the ten bestselling toys were linked to children's media.[31]

Media and toy companies that profit hugely from licensed characters have a vested interest in preventing children's creative play—and stifling their creativity. Kids who are creative need fewer toys because they can figure out different things to do with the ones they have. Toys that promote creativity are less likely to be huge moneymakers because they can be used repeatedly in lots of different ways. The big money in toys is not in selling families a particular toy or a one-off game, film, or TV program. In addition to creating a brand icon that can be licensed to companies making

---

*These program-length commercials had been banned in the 1970s.

other products, the big money is in selling kids on the necessity of acquiring a series of toys, apps, and games and convincing kids that they always need the latest one.

One classic brand that has done a brilliant job of mastering this concept is Pokémon. The brand's theme song features a pulsing chorus of "Gotta catch 'em all." Winning—good triumphing over evil—can only be achieved by capturing all the Pokémon characters. For children, that means collecting every one of the Pokémon characters. Pokémon is kind of like the toy version of Lay's potato chips with its iconic tag line "Betcha can't just eat one." Perhaps some children are satisfied with owning only one Pokémon character, but I've never met them. I understand the appeal of Pokémon—the characters are incredibly cute, and the battles are exciting. I also know children who enjoy drawing them, and kids who use them to connect with friends. The company's business model, however, is both brilliant and diabolical. It reminds me of Tantalus, the figure in Greek mythology who is doomed to stand in a pool of water under a fruit tree, forever unable to drink or eat from either, rendering him perpetually hungry and thirsty. Pokémon's perpetual profits come from inculcating longing in children that can only be fulfilled by completing the set. But by adding new characters, or iterations of characters, to their product line, the brand ensures that the set can never be completed. There's always another character to buy.

It's not that I think that the CEOs of huge toy companies sit around plotting to make children and their parents miserable. It's that their nearly singular focus on generating profits leads them to manufacture toys with enough electronic bells and whistles that they market well in fifteen-second spots or whose popularity rides on the coattails of popular media properties or that are unsatisfying and incomplete in and of themselves because they are part of a set.

Traditionally, for kids having what Winnicott might call "good enough" childhoods, the adults with whom they mostly interacted, and who had the most influence over them, were parents, other caregivers, and teachers. They were all familiar family or community members who were at least supposed to have children's interests at heart. For better or worse, it's common to raise children to be wary of strangers. Yet in a digitized, commercialized culture, we blithely turn vast portions of a child's day over to strangers. We don't see these people. Our kids never meet them. But these strangers know an enormous amount about our children. They know how to capture their attention, to exploit their vulnerabilities, and to trigger their longings. These are the strangers who own, manufacture, and advertise the apps, toys, and games that occupy children's time and whose jobs demand that they develop and market products that generate big bucks regardless of their impact on the kids who use them.

Let me be clear. There are lots of toys that are wonderful for kids to play with and that enhance their lives. There are toymakers who care a great deal about children's creative play. Except for those aimed at babies and toddlers, there are also media programs and apps that contribute to children's healthy development (although it's increasingly hard to find ones that don't simultaneously exploit children by targeting them with ads). The real problem is that the prevailing business model for tech, media, and mega toy companies fosters design and marketing decisions that prioritize profits—even at the expense of children's wellbeing.

2

Who Wins the Games Tech Plays?

Tech isn't morally good or bad until it's wielded by the corporations that fashion it for mass consumption.
—ADAM ALTER, *Irresistible: The Rise of Addictive Technology and the Business of Keeping Us Hooked*

My puppet Audrey Duck and I are chatting with three-year-old Hazel just before her bedtime. She dashes around her bedroom, introducing us to her stuffed Humpty Dumpty, offering Audrey a cup of tea, and showing off her wooden duck. We sing what she calls "Blah Blah Black Sheep," her version of that old nursery rhyme, and other songs, and we laugh together at the silliness we're creating. It's like hundreds of other play experiences I've had with children—except that Hazel is thousands of miles away.

Thanks to the wonders of digital technologies, during the long COVID lockdown Audrey and I got to play with kids who were cooped up at home. I'm grateful for the digital platforms that made these sessions possible, but I'm also acutely aware of both the irony and the limits of my gratitude. After all, I've long been critical of the ways tech companies exploit kids—from incessantly hawking in-app purchases or costly upgrades to invading children's privacy to equipping apps with features that make it hard for kids (and adults) to tear themselves away.

So even if we and the children we love rely on tech to meet some of our basic needs, it's essential to remember that the prime directive of tech companies is to generate profit, not to provide a public service. The methods and techniques they use to rack up huge profits, along with the business models that made and continue to grow the fortunes of three of the four wealthiest people on the planet,[1] are inherently harmful to all of us—particularly children whose immature brains and lack of experience make them even more vulnerable to manipulation than adults are.

Video chat platforms and social networks may seem like digital versions of a public square—but they aren't. The key word is "public." No individual or no corporation owns public space. No one profits from its use. No one calculates how best to use our every word, our interests, our friends, and our longings to persuade us to buy things. And, certainly, in a public space, no one exploits our vulnerabilities or our basic human needs to keep us hanging out long past the time when it's in our best interest to move along.

Given that most contemporary marketing to children is delivered by increasingly sophisticated technologies, efforts to curb advertising to kids often focus on limiting tech and media companies' access to children. That's why I and other advocates working to increase children's tech-free time are often labeled Luddites, an epithet erroneously used interchangeably with technophobe. Dictionaries tend to define "technophobe" as someone who either fears or dislikes new technologies and can't use them with confidence.[2] Given that most of my work life has been spent in front of one kind of screen or another, I don't think I qualify as a technophobe. I do, however, sympathize with actual Luddites, who are given a bad rap. In the early nineteenth century, displaced textile workers in England were forced into abject poverty and even starvation when mill owners replaced them with machines that worked faster, more efficiently, and at less expense than living, breathing workers did. With no legal recourse available to them, many of these workers acted against their former employers by damaging, and even destroying, the machines that replaced them. It's not that these workers, the Luddites, were necessarily categorically opposed to machines—rather they were opposed to the horrific effects the use of machines was having on their lives and the lives of their families.

I suspect that if the mill owners had fairly compensated their workers or considered and acted on the ethics of replacing them with machines, the Luddite movement would not have existed. In other words, it wasn't the machines per se that the Luddites were protesting, it was the business practices of the mill owners who installed those machines.

In that sense, I identify with the Luddites. It's not the technologies, or devices themselves, that I object to, it's the business practices of tech and media companies who own them. And while I do not advocate going around smashing iPads, smartphones, Chromebooks, or any digital machine, I do advocate for limiting the power of tech and media companies to profit from exploiting and potentially harming children.

Over the years, I've thought a lot about those nineteenth-century British textile workers raging against the machines that took their jobs. The Luddites' plight is emblematic of a central problem with all kinds of consumption-enhancing inventions, including digital technologies. These may be wondrous creations that generate great wealth for the companies that claim them. Consumer-related tech advances, like the smartphones, tablets, and social networks, are celebrated as progress and widely embraced, often with little public discourse about the social, emotional, psychological, and ethical consequences of how they're used.

Technologies are problematic when they optimize profits at the expense of the health and wellbeing of individuals and the larger society. Yet no independent review of the potential harms and benefits of tech products is required before they go on the market. As a result, advocates for children's wellbeing in a commercialized world constantly must play catch-up.

I can think of only a few incidents when advocates had enough lead time before the launch of any new product—tech or not—to stop its production. In 2017, for instance, Fairplay helped prevent Mattel from launching Aristotle, a smart device designed to be installed in children's bedrooms from birth till the late teens and marketed as "an all-in-one nursery necessity."[3] As described in the *New York Times*, Aristotle had many of the characteristics of a wonderful caregiver, including "the ability to soothe a crying baby, teach ABC's, reinforce good manners, play interactive games, and help kids with homework."[4] In other words, Mattel was about to launch a device designed to serve as a baby's major comforter, teacher, civilizer, and sometime playmate—a surrogate parent.

In the same article, Josh Golin, who succeeded me as Fairplay's executive director, pointed out the obvious concerns about privacy, noting that "when you have a device with a camera and a microphone that's going to be in young children's bedrooms, there is the potential to collect so much data on children that can be used and shared with advertisers and retailers."[5] That's bad enough, but his second point is even more troubling: "Then there are all these child development concerns about replacing essential parenting functions with a device."[6]

The child development concerns he raised are huge. They have to do with the crucial process of attachment—the lifelong bond that develops between parents and children that lays the foundation for social, emotional, and psychological wellbeing. That's why the parental functions Mattel attributed to Aristotle, especially its alleged capacity to comfort a crying baby, catapulted me back to reading in an undergraduate psychology class about Harry Harlow's nightmarish experiments on baby monkeys.

To discover whether the need for attachment is innate, Harlow removed newborn monkeys from their mothers to see if they would bond with inanimate objects—the most successful of which was a wire construction covered with soft cloth and equipped with a fake face. In a way, the monkeys did bond to these objects, in that they consistently turned to them for comfort and a sense of security.[7] As the baby monkeys in his lab grew up, however, Harlow's research showed that babies "raised" by these objects became significantly dysfunctional adults.

The phase of Harlow's attachment experiment that haunts me to this day for its cruelty is when, after the baby monkeys attached to an inanimate "mother," the "mother" became abusive. The objects might stab the babies, for instance, or blast them with cold air, or violently shake them. The babies remained attached and persisted in seeking comfort and security from these inanimate objects, no matter what horrors the "mother" inflicted on them.[8]

I'm relieved that Harlow's horrific mistreatment of the monkeys in his lab would not be allowed today. The irony is that his findings served to illustrate the central importance of attachment—the bond between infants and parents that in "good enough" families is strengthened as children grow and develop. By replacing crucial parental functions with a machine, Mattel was potentially disrupting the development of healthy attachment and potentiating babies' emotional dependency on Aristotle, and by extension on the corporation itself. Robb Fujioka, Mattel's chief products officer, was upfront about Mattel's hope that children would bond with Aristotle. If their product launch was successful, kids would "form some emotional ties" to the device,[9] and children would grow up being "nurtured" by an entity whose interest in them is primarily financial. And just as Harlow's goals had nothing to do with the wellbeing of baby monkeys, Mattel's prime directive has nothing to do with the wellbeing of baby humans. Like Harlow's structures, Aristotle could harm children—not by stabbing or shaking but by depriving babies of crucial bonding time with their actual caregivers, by fostering dependence on Aristotle for nurturing and companionship, and by shaping children's desire for the things Mattel, and its corporate partners, including junk food producers and media companies, manufacture and market.

Advocates were successful in their efforts to stop Aristotle's release in part because there was—unusually—ample time between when news of its impending debut went public and the product's actual launch. Mattel announced its plans for the device in January 2017.[10] In May, Fairplay and the Story of Stuff Project launched a petition that garnered more than twenty thousand signatures urging Mattel to scrap the project because, among other reasons, babies and older children shouldn't be trained to bond with "data collecting devices."[11] In September 2017, Senator Ed Markey (D-MA) and Representative Joe Barton (R-TX) sent an open letter to Mattel citing concerns that Aristotle had "the potential to raise serious privacy concerns as Mattel can build an in-depth profile of children and their family."[12] The next month, Mattel scuttled its plans for Aristotle, telling the *New York Times* that the company's new chief technology officer "conducted an extensive review of the Aristotle product and decided that it did not fully align with Mattel's new technology strategy."[13]

It's terrific that Aristotle never made it to market, but it's also a significant problem that there are no rules, laws, or conventions in place at this time to prevent other companies from launching a similar product.

It doesn't have to be this way. A society that values public health over profits could invoke the precautionary principle to prevent companies from marketing potentially harmful products and practices. The precautionary principle is a guide to societal decision-making rooted in that old adage "look before you leap." It acknowledges that innovation in science and technology proceeds faster than does an understanding of the ecological, public health, and humanitarian consequences of the inventions it spawns. Employing the precautionary principle allows science-based public health concerns to influence how or whether a product or practice can be used. The precautionary principle has been officially adopted by the European Union and various international agreements, but by not the United States.[14]

The United Nations Education, Science and Culture Organization (UNESCO) recommends invoking the precautionary principle "when human activities may lead to morally unacceptable harm that is scientifically plausible but uncertain" and advises that "actions shall be taken to avoid or diminish that harm."[15] UNESCO identifies as "morally unacceptable" harm that threatens human life or health, is irreversible, is in potential violation of human rights, or is unfair—including harm to future generations.

The precautionary principle has been invoked primarily to prevent environmental degradation or physical harm to humans. Suppose, in the face of credible but not conclusive scientific evidence to the contrary, media, tech, and marketing companies that target children had to provide research demonstrating that their products and practices do no harm. As it is, in the United States, even when there is plausible evidence suggesting harm to kids, media and tech products continue to target kids. For instance, despite decades of research suggesting that playing violent video games is a risk factor in aggression,[16] these games are still marketed to children and teens.[17] So even with years of research that suggests harm, the burden of proof remains on advocates to prove harm definitively instead of on companies to prove lack of harm.

If children's exposure to digital technologies was only minimal, or if companies were subject to effective government regulations designed to prevent them from exploiting children, then the industry's laser focus on profit wouldn't matter so much. Even before the pandemic, children—including babies and toddlers—spent enormous amounts of time with tech. Pre-COVID, on average, babies spent 49 minutes a day with screens. The amount of time tripled when children became preschoolers, to averaging 2.5 hours per day. For five- to eight-year-olds, screen time averaged more than 3 hours per day.[18]

On average, pre-pandemic, eight- to twelve-year-olds used entertainment screen media for 4 hours and 44 minutes each day, and teenagers used it for 7 hours and 22 minutes a day.[19] That these are averages means some kids spent significantly less time with screens each day and some spent significantly more. For instance, almost 25 percent of children between the ages of zero to eight spent no time with screens, and an equal percentage spent more than 4 hours per day.[20] Kids from households with lower incomes spend nearly 2 hours more with screens than kids from higher-income households. In addition, Black and Latinx children spend significantly more time on screens than their white peers.[21]

Once the pandemic hit, children's screen time increased as a result of school closures, physical-distancing requirements, and parents' desperate efforts to keep their children occupied while they worked from home.[22] Josh Golin observed in his recent testimony to Congress:

> COVID-19 has accelerated these trends, and screen time for children is estimated to have increased by 50 percent during the pandemic. During the same period, online messages sent and received by children increased by an incredible 144 percent. Thirty-five percent of parents report that during the pandemic they allowed their children to begin using social media at a younger age than they originally planned.[23]

Children's time with digital devices has now supplanted their time with television. But what are they doing on their devices? Primarily, they are watching online videos,[24] which is ironic given that when the tech industry pushed phones and tablets as beneficial for kids, they argued that kids would be active on these devices rather than passively watching a screen.[25] In fact, watching videos and television accounts for almost 75 percent of all the time kids zero to eight years old spend with screens. Gaming accounts for 16 percent, while reading, homework, and video chatting account for only for only 5 percent.[26]

Before I say any more about the tech industry and its impact on children, I want to acknowledge there is a huge digital divide—and it's problematic. It's true that in developed countries, smartphones provide almost universal access to the internet, but as the pandemic made painfully clear, phones don't readily lend themselves to completing school assignments or participating in distance learning. The eighteen or so months between March 2020 and September 2021, when so many schools were shut down and so much of schooling took place online, made clear that children without high-speed internet access or whose family didn't own larger devices like laptops or tablets were at a serious disadvantage.

Research suggests, however, that just as there is a digital divide, there are also divides that are at least exacerbated by the marketing myth that tech is an adequate educational substitute for human interactions and hands-on experiences with the world.

Take language acquisition, for instance. By the time children begin kindergarten, there are huge discrepancies in the number of words they know. The disparity in children's language acquisition appears to be present as early as eighteen months and can affect how well kids do in school—including in fundamentals like math and literacy.[27]

What makes the difference? Language acquisition is inherently social.[28] From infancy, children's vocabulary and verbal fluency is linked to how much exposure they have to language. For babies to learn language, however—and this is important—it has to be generated by people, not by machines.[29] That's why the American Speech-Language-Hearing Association encourages parents and caregivers to talk, read, and sing to and with children starting from infancy.[30] That's also why our own tech use around infants and toddlers can interfere with language acquisition. When parents or caregivers are on smartphones, they speak less to their babies.[31] And, particularly for infants and toddlers, more time with screens of all kinds is associated with diminished language development,[32] even when the apps or videos in use claim to build vocabulary.

In actuality, scant evidence exists that using technologies to teach babies language, or much of anything else, is effective. The research I've seen that even remotely makes the case for the educational benefits of screen usage by babies focuses on screen use in the company of parents or caregivers who reinforce the content. The problem, of course, is that parents tend to use screens to occupy babies so they (the adults) can do other things.[33] There is no credible evidence that watching screens or using devices on their own is educational or meaningfully beneficial for babies.[34] In fact, research suggests that it may be harmful.[35]

Everything we know about how our youngest children learn points away from screens to what they do naturally—engage with the people who love them best and explore the world around them with all five of their senses. Research that explored the impact of television on children suggests that the more time young children spend in front of TV, the less time they spend engaging in two activities proven to be beneficial: interacting with parents away from screens and spending time actively involved in creative play.[36] For every hour preschoolers spend watching a screen, they spend 45 minutes less in creative play. Babies and toddlers lose even more time in creative play than their older brothers and sisters lose— 52 minutes for every hour of TV.[37]

While most of the research on babies and screens is still mainly about television and videos, it appears that the use of touch screens may also be problematic. In Britain, one longitudinal study indicates that babies and toddlers who spend a lot of time with touch screens appear to be more easily distracted as preschoolers and to have a harder time controlling their attention.[38] Research also suggests that heavy media use at the age of two is associated in preschoolers with difficulties with self-regulation, which the researchers define as the ability to control "behavior, emotional reactions, and social interactions despite contrary impulses and distractions."[39]

What's especially concerning about encouraging babies and toddlers to spend time with screens is that they are likely to spend more time with screens when they are older.[40] That's particularly worrying because research suggests that children's excessive screen use is harmful at every age. Across ages, hours spent with screens are linked to a host of problems, including, but not limited to, childhood obesity, sleep disturbances, depression, and doing less well in school.[41]

There's no question that, in moderation, once kids outgrow toddlerhood, quality educational television programs and apps can be beneficial. They can encourage, for instance, prosocial qualities like empathy and may contribute to mitigating racial and ethnic biases.[42] TV programs and apps can both teach[43] and reinforce academic skills.[44]

So what does moderation mean? The American Academy of Pediatrics' evidence-based recommendations for young children and technology recommends no screen time (except video chatting) for babies up to eighteen months; minimal screen time, and only with an adult, for toddlers eighteen months to two years; and no more than one hour a day of entertainment screen media for preschoolers.[45]

Moderation, however, is where the incompatibility between prioritizing profit and prioritizing public health is most blatant. Tech and media companies flourish by capturing our attention, keeping our attention as long as possible, and doing everything they can to lure us back for more. It's in these companies' financial interest to promote excessive screen use. Just as casinos profit from addiction to gambling, companies that produce apps, games, and social media also profit from addiction to the content they produce.[46]

As psychologist Jean M. Twenge, author of *iGen*, said, "Overuse of digital media is linked to unhappiness, depression, and suicide risk factors, while limited use appears to be low-risk. However, many apps, games, and websites make more money the more time children and teens spend on them. Many companies are doing everything they can to make sure that limited use is not the norm, and that needs to stop."[47]

Tech companies certainly aren't going to stop voluntarily, especially those whose business model depends on inculcating powerful, habitual ties, if not addiction, to devices. As adults, we are certainly susceptible to becoming enmeshed in, and in thrall to, the digital world. And children, with their developing and adaptable brains, are particularly vulnerable to capture. The first step in disentangling ourselves from digital media is understanding the role we and our children play in generating huge profits for companies like Meta, Google, and Microsoft.

For those of us whose strengths and interests lie outside of the world of math, science, and engineering, I find there is an aura of mystery surrounding digital technologies that acts as a barrier to understanding exactly how tech companies make money. These devices we love and use daily operate so seamlessly and seductively that they seem like magic. Wireless technology, especially, seems magical since we can't see how the images, emails, and information we pull up on our screens arrive there.*

At the same time, we know that tech isn't magic. We know that technology is harnessed by people, unlike many of us, who are fluent in math, science, and engineering. That much of the innovation around technology comes out of elite universities like MIT and Stanford makes even attempting to understand it seem more daunting. That's why I've come to believe that a necessary first step toward making conscious, responsible, ethical decisions about how we as governments, communities, families, and individuals manage the tech industry's access to children is acquiring at least a basic understanding of how the mechanisms that keep us glued to our devices and generate huge profits function. I found that my own understanding was predicated on deciphering terms that were previously unfamiliar to me, so I've defined them in the box on pages 45 and 46.

Tech conglomerates generate huge profits and wield enormous influence by offering us convenience, distraction, information, and connections. In exchange, often without realizing it, we relinquish

---

*Special thanks to Criscillia Benford, who first described to me the ambience surrounding new digital technologies as "magical."

Tech Jargon in Lay Language

Algorithms are like recipes written by programmers for computers. They are sets of equations and rules that, when followed, allow computers to complete assigned tasks.

Machine learning algorithms make predictions or classifications by inventing "learning" rules from stored data that was fed to them or from data extracted from real-time human/computer interactions.

Predictive algorithms are machine learning algorithms that use previously collected data to "learn" to predict something that has yet to happen. Research has shown that the rules these algorithms invent to predict the future often reproduce biases in their training data.

Internet of things (IoT) is a catchall phrase for the billions of smart devices—toys, microwaves, vacuum cleaners, thermostats, toothbrushes, etc.—with internet connectivity. While some of the data these devices capture through their sensors is useful to device owners, most of it becomes monetizable by device providers once it is uploaded to a cloud-based platform like Salesforce IoT.

User experience (UX) design molds how users behave with and feel about interactive products like smartphones, tablets, apps, websites, and IoT objects and devices.

Persuasive design is a subset of UX design practice inspired by principles from behavioral psychology to persuade users to adopt *target attitudes* and/or perform *target behaviors* (e.g., stay on the site, click, like, pin, swipe, post, join, comment, subscribe, message, buy).

(continued on next page)

> **Variable reward schedules** refer to a common tactic for inducing target behaviors in users by offering rewards, which may differ widely in value, on an unpredictable schedule.
>
> **Behavioral advertising** (or behavioral targeting or data-driven advertising) is advertising based on data collected from peoples' online behavior and aggregated with data from individuals with similar traits or behaviors.
>
> **Nagware** consists of pop-up ads often embedded in "free" apps urging users to buy an upgrade or a premium version of the app. They often appear when apps are open and then may "pop-up" again after a set amount of time.
>
> **Push notifications** are notices designed to encourage and prolong the time we spend with our devices. Types of push notifications include news of what "friends" are doing on social media, "likes" garnered by posts, or alerts about updates to a particular app.
>
> **Infinite scroll** is just what it sounds like. It's a design technique to prolong engagement by loading content continually as users scroll down a page.
>
> ---
>
> Copyright Criscillia Benford. Adapted with permission.

our privacy in the form of our personal information, time, and attention, as well as our cash. Companies like Amazon and Meta exploit the science of persuasion—what is known about influencing human behavior—in the service of persistently nudging our use of their products and services to become habits, which can ultimately become addictions.

One particularly powerful technique to keep us hooked is the use of intermittent rewards. We don't just post on social networks, we check our offerings repeatedly for "likes," "shares," and comments. For young children, intermittent rewards come in the form of virtual stars, points, or prizes. It turns out that attaining rewards some of the time, but not all the time, is a powerful motivator. Just as slot machines are programmed to let gamblers win intermittently, tech companies also provide intermittent rewards—in the form of those "likes" on social media or leveling up and virtual prizes in online games—to keep us constantly engaged. Each digital "reward" triggers a little squirt of dopamine, a powerful neurotransmitter that triggers the seductive combination of enjoyment or excitement and desire.[48]

Once captured in this loop of pleasure and longing, we and our children are subject to unprecedented surveillance. At minimum, most of our behavior online is tracked. Increasingly, our behavior offline is tracked as well through location devices on our phones and through the Internet of Things (IoT), a catchall phrase describing "smart" toys, appliances, watches, and more. More and more, children are vulnerable to surveillance when they play. Sales of smart toys, including the physical toys and their associated apps, are expected to reach almost $70 billion in 2026.[49]

Tech companies call the process of continually collecting information about us as we use their offerings "data mining." Because this process involves real-time surveillance of how we live our lives, the fact that companies are mining us and our children seems particularly creepy and reminiscent of Orwellian science fiction. And it is. But the practice of mining potential consumers isn't new.

As I wrote in *Consuming Kids*, a 2001 article in *Brand Strategy* urged companies targeting kids to engage in "relationship mining," which the author defines as "an overall description of the method for understanding family forces. Mining refers to the process of uncovering the motivations of different family members and the reasons for particular outcomes when conflicting needs occur."[50] For those of us who free-associate, the metaphor of mining family relationships is particularly and painfully evocative. Families are perceived as a repository (the mine) containing valuable raw materials available for extracting—and exploiting.

As with so much of marketing to children today, the specific intent to exploit them hasn't changed over the years, but the tools for doing so are increasingly precise, invasive, powerful, and seductive. To extend the mining metaphor, the information tech companies continually collect from us serves as the raw material for predictive algorithms that calculate how we and others with similar behaviors or characteristics are most likely to respond to information, products, or prompts.

What the algorithms don't consider is what's best for children. Take the endless supply of "recommended" videos that follow one after the other on YouTube. Kids who start out watching clips from the benignly popular British animated series *Peppa Pig*, for instance, on YouTube, and continue to watch each recommended video as it appears on their screen can find themselves immersed in violent, sexually suggestive, and drug-related content.[51] But regardless of the suitability of the content children are led to, the fact that YouTube knows enough about young viewers' interests to seduce them into extending their viewing is troubling, especially when excessive screen time is so pervasive and problematic.

Guillaume Chaslot, a former YouTube engineer, puts it more strongly: "Recommendations are designed to optimize watch time, there is no reason that it shows content that is actually good for kids. It might sometimes, but if it does it is coincidence. . . . Working at YouTube on recommendations, I felt I was the bad guy in *Pinocchio*: showing kids a colourful and fun world, but actually turning them into donkeys to maximise revenue."[52]

Chaslot left YouTube and founded AlgoTransparency, an organization that works to shed light on the influence of algorithms on deciding what we see when we're online. Today, data gathered from our past behaviors, online and off, is used to create predictive algorithms that tech companies sell to advertisers as tools for targeting specific ads to people who are most likely to respond to them. Habits, strengths, loves, hates, and desires as they play out online (to say nothing of our age, location, gender, and socioeconomic status) are the raw materials mined by tech companies to keep us glued to websites, games, and platforms.

Kids frequently experience this kind of surveillance and monetization in popular online games. Eleven-year-old Mark was thrilled when I asked his parents if I could watch him play the wildly popular multiplayer game Fortnite. He put on his headset, logged on with a friend, and was instantly lost to it—shooting down funny-looking creatures and plotting real-time strategies with his partner in crime.

Fortnite is what's called a "cooperative sandbox game," which offers players opportunities to play creatively with others. As someone who has fun playing games of all kinds, I get the appeal. Fortnite—and games like it—offer excitement, immediate gratification, and opportunities to hang out with friends. But there's a big difference between playing these online sandbox games and playing games in a real sandbox.

When kids connect with each other on Fortnite, or other sandbox games, their play is continually monetized by companies exploiting their vulnerabilities by fomenting envy and longings to belong. It's true that children building actual forts and castles in a sandbox may experience envy for other children's toys or they may experience being excluded. But physical sandboxes are normally free of adults whose job it is to purposely manipulate kids into situations that trigger those experiences.

Games like Fortnite monetize children's play by constructing virtual worlds rife with in-your-face conspicuous consumption, complete with clearly visible haves and have-nots. They sell status symbols in the forms of virtual accoutrements for avatars (called *skins* in Fortnite-speak) and dollops of silliness like weird made-up dances (called *emotes*) thrown in. Players pay to customize their avatar's costume, colors, tools, and weapons. Fortnite also generates cash by charging $10 quarterly for a "Battle Pass" that allows exclusive access to cosmetic updates like more complex skins, new characters, pets, and more.[53] Kids playing Fortnite with friends are constantly reminded who has money to spend and who doesn't. It works. In 2018, the company reportedly raked in $300 million per month from in-app purchases, which comes to $3.6 billion for the year.[54] By 2020, revenue was down but still a respectable $2.5 billion.[55]

Mark's father worried about how much money Mark was spending on virtual Fortnite gear. "It's weird," Mark's father told me. "These days he wants to spend more money on things that don't actually exist than on real stuff." At eleven, Mark had his own money primarily from saving his weekly allowance. And once kids have their own money, it gets harder for parents to control what they spend it on. Of course, Fortnite isn't the only game to entice kids into spending money on virtual products, but it's one of the most popular and pernicious. And Mark's father had one more concern. "I can't get him to stop without a fight," he said. I hear that complaint a lot from parents struggling to get their kids to stop playing video games or to get off social media.

Mark's difficulty disengaging from Fortnite is no accident. Nor is it an accident that most of us are in thrall to (and enthralled with) our devices. Tech companies have us where they want us. And not only do they possess the tools to keep us there, we participate in ensuring their success. The cliché that comes to mind as I write this is "digging our own graves." Here's how it works. The algorithms driving apps and social media sites continually hone their ability to persuade us to increase the amount of time we spend online. The algorithms evolve based on what they learn about us. The more we're online, the more information about ourselves we provide to the algorithms. The more information we provide, the better the algorithms "know" us and the more effective they are at keeping us hooked. If it's hard for adults to disengage, imagine what it's like for children, who have a much harder time with self-control and discipline and whose judgment is still immature.

Bolstered by progressively sophisticated technology and

increased understanding of human behavior, today's marketing techniques are the most powerful and persuasive to date. Yet the extraordinary power of today's marketing didn't spring fully formed out of nowhere. Rather, it is one additional rung in the ever-evolving and lucrative marriage of technology and psychology. Digital marketing is merely one step up in advertising's long history of combining advances in technology and research on human psychology to capture our attention for the sole purpose of selling us stuff. Tech companies routinely employ psychologists to increase their power to capture and persuade.[56]

It's a sad fact that psychologists have long brought their vast and varied knowledge of human emotions, thinking, and behavior to advertising. And it's long been the role of engineers to apply the latest material science to expand the reach of advertising and to deliver its messages in new and exciting ways.[57] As the sciences of psychology and technology evolve in combination, the power and reach of commercialism expands, including the practice of manipulating children for profit. For decades, psychologists have routinely helped companies market successfully to children by employing the principles and practices of child psychology—from developmental theory to diagnostic techniques—for the sole purpose of increasing corporate profits.

Developmental psychology—the study of the cognitive, emotional, and social changes children go through as they grow up—traditionally provides the foundation for all kinds of public policies designed to protect children and to promote their wellbeing. That's why it troubles me that some psychologists—whose work is supposed to benefit humanity—provide corporations with increasing power to influence anyone. Today, developmental psychology is just another tool for what the advertising industry calls "market segmentation" or "target marketing."[58]

Way back in 1999, I joined a group of psychologists urging the American Psychological Association (APA) to declare that it was unethical for psychologists to work with marketers targeting children. APA refused, although they did appoint a group of psychologists, including myself, to a task force whose job was to investigate the problem, write a report, and make recommendations.[59] Based on available research, we recommended that APA support restricting advertising and marketing to kids under the age of eight and restrict marketing in schools. APA accepted our recommendations as policy.[60]

Almost twenty years later, I joined another group of psychologists in a letter urging APA to declare that it is unethical for psychologists to work with tech companies to manipulate children.[61] As Richard Freed, one of the organizers of that effort, explained, "The destructive forces of psychology deployed by the tech industry are making a greater impact on kids than the positive uses of psychology by mental health providers and child advocates. Put plainly, the science of psychology is hurting kids more than helping them."[62] As I write this, the APA has taken no action.

Meanwhile, there's hardly anywhere in the world of tech and media for children that is truly commercial-free. When public television and radio began, they were supposed to be free of advertising and marketing. Over the years, however, severe cuts in funding to the Corporation for Public Broadcasting (CPB) forced programs on the Public Broadcasting Service (PBS) to rely heavily on generating their own funding. As a result, even programs with wonderful educational content rely on commercial sponsorships, brand licensing, and apps and games to keep kids' attention. Today, there is just about nowhere kids can travel in the digital landscape without being targeted with advertising. In addition, as I discovered at a Clinton Foundation meeting in Chicago a few years ago, even executives who produce public media for kids are reluctant to take a public stand on the importance of limiting young children's exposure to television and digital devices.

I was excited about attending this meeting because it included spending private time with a group of decision makers from PBS, CPB, Sesame Workshop, and more. At that time, the American

Academy of Pediatrics (AAP) recommended discouraging screen time for children under two and limiting entertainment screen media for older kids to no more than one to two hours a day.* Wouldn't it be great, I thought, if the producers of public media for young children, who are *supposed* to put children's best interests first, came together to recommend that parents follow the AAP recommendations? They refused—at least publicly.

A couple of my fellow participants told me privately that they agreed with the limit. But it seemed they couldn't risk sending that message to parents for fear they would lose corporate sponsors and licensing partners. It's the same reasoning used by purveyors of commercial media for kids, which underscores the sad truth that while underfunded public media available for kids today produces some wonderful programs, and may be more educational than most commercial media, it is financed through a similar business model, at the expense of children's wellbeing.

In the world of for-profit apps and video games, the sandbox game Minecraft epitomizes tech's potential to offer children wonderfully fun, creative, and educational experiences even as its prevailing business model exploits and corrupts those same experiences. I first encountered Minecraft in March 2014, when I was preparing to participate in a panel about tech, which included a teacher who used the game in his classroom. I was charmed by it. Six months later, Microsoft bought Minecraft from its creator for $2.5 billion.[63]

What excited me most about Minecraft was that it was pretty much completely unstructured. At the time, all the game consisted of was dropping players off in virtual Lego-like landscapes where they had to rely entirely on their own creativity and problem-solving abilities to survive. You could play alone or collaborate

---

*The 2013 recommendations were still in effect at the time: Council on Communications and Media, "Children, Adolescents, and the Media," *Pediatrics* 132, no. 5 (2013), 958–61.

with friends online. At that point, Minecraft was by no means commercial-free—there were ever-increasing numbers of licensed products,[64] including T-shirts and even a Lego set.[65] I was bothered by the commercialism—which of course now seems rather quaint—but I was still intrigued by the game.

Like Mark, ten-year-old Henry was thrilled to show me what Microsoft's Minecraft is like today. He was even more thrilled when his dad said, "Let me show you what I *don't* let Henry play on Minecraft," and we were transported to a virtual multiplayer battlefield, which still looks like Minecraft—blocky landscape, blocky avatars, and all. But it's now an awful lot like Fortnite. Minecraft replicates Fortnite's class system of haves and have-nots by touting in-game virtual add-ons, including skins, pets, weapons, and more.[66] While his dad and I chatted about Minecraft, Henry spent a happy fifteen minutes fighting his opponents and blowing things up. The excitement offered by this version of the game comes more from opportunities for destruction than construction—which is why he is normally not allowed to play it.

Microsoft's über-commercialization of Minecraft is distressing but not unexpected. Yet, even before that happened, I had a weird exchange with a Minecraft enthusiast. At that panel discussion in 2014, I found myself treading rather lightly on Minecraft when the teacher, my fellow panelist, extolled its virtues. But then, with great excitement, he described how his daughter had built a treehouse on Minecraft. "We live in New York City," he enthused, "She never built a treehouse before." That was too much for me. "She *still* hasn't built one!" I exclaimed. It's a conflation that worries me.

In an op-ed in the *New York Times*, the psychologist and tech-critic Sherry Turkle describes the way technology insinuates itself into our lives as going from "better than nothing to better than anything." And then she adds, "These are stations on our voyage to forgetting what it means to be human."[67] Turkle's concerns go way beyond apps and video games to a time envisioned by futurists when robots replace humans as, for instance, caretakers for the elderly, children, and the infirm. Where we no longer distinguish between machines that simulate empathy and humans who actually feel it. My own glimpse of that future goes back to that 2014 panel and that teacher who couldn't—or wouldn't—distinguish between constructing a virtual treehouse in Minecraft and the experience of building one outside, perched way up in a tree, pounding nails into boards. Our interactions with tech may be compelling, exciting, illuminating, meaningful, and fun. But, to echo Sherry Turkle, it's a mistake to buy into the tech industry's tendency to equate them with our interactions with the actual world or what, in tech-speak is called "IRL," or "in real life."

When COVID-19 first upended the planet, digital technologies became increasingly key to helping those of us with access to them protect and nurture the physical and mental health of our children and ourselves. Our experience seemed to prove what tech and media companies have long argued—that their offerings are essential to modern life. If that's the case, however, we need to change their business model. Unlike other essential services, such as utilities for instance, the tech industry vies for our attention, invades our privacy and our children's, and monetizes our personal information by subjecting us to inescapable, data-driven advertising. We need to regulate Big Tech, especially in how companies target children. And, while we marvel at the amazing intelligence of the machines that are here, and of those to come, we need to value the ways we're different from them—and celebrate the best of what it means to be human.

3

And the Brand Plays On

What I would love is to have any boy in the world who thinks of pirates to think of . . . Disney pirates.
—Bob Iger, former CEO and current executive chairman and chairman of the board, Walt Disney Company

I am sitting in the sparsely populated state-of-the-art auditorium at the San Francisco Jazz Center. I'm attending PlayCon 2018, a Toy Industry conference billed as the "biannual gathering of thought leaders forming the future of our ever-changing, and at times, challenging industry."[1] Challenging is right. Two months earlier Toys "R" Us, the iconic big-box toy chain that accounted for 15 to 20 percent of domestic toy revenue,[2] had filed for bankruptcy and was closing. At its height, Toys "R" Us operated nine hundred stores just in the United States. Its mascot, Geoffrey the Giraffe, was as recognizable as Kellogg's Tony the Tiger.[3] Worry about the ramifications of what was the last of the giant brick-and-mortar toy chains was palpable, even in the conference schedule. The day began with an open forum focused on surviving bankruptcy and "industry disruption."

PlayCon attendees include executives from Hasbro and Mattel, two of the largest toy conglomerates in the world, both of which blamed Toys "R" Us, at least in part, for significant losses in revenue.[4] As a result, Mattel had cut 2,200 jobs—22 percent of its non-manufacturing workforce worldwide. For the first time in fourteen years, Lego—not among the attendees—had also lost revenue, particularly in North America, and cut 8 percent of its workforce.[5]

There's no question that the demise of Toys "R" Us at least temporarily harmed toy manufacturers, most of whose companies, large and small, depended on the stores for a significant portion of their sales. It was certainly a major tragedy for the families of more than thirty thousand employees who were initially dismissed with no severance pay.[6] (After months of protest, some, but not all, of the workers who lost their jobs were compensated.)[7] The rise and fall of the last big-box toy chain may seem irrelevant today, but I find myself thinking about the company as I listen to presentations by the toy and tech executives at PlayCon. Toys "R" Us devised extraordinarily effective marketing strategies that lay the groundwork for companies that target children today. As one market researcher said in the *Wall Street Journal* during the company's heyday: "Its strength isn't in its product but the way they sell it."[8]

It turns out that Toys "R" Us was down in 2018, but apparently not out. It will be opening inside brick-and-mortar Macy's locations in 2022 and, thanks to its deal with the department store chain, it now has an online presence.[9] But I'm not jumping up and down about its return. Toys "R" Us advertised heavily, and brilliantly, to kids—positioning its stores as fun-filled, kid-friendly, *necessary* destinations for families. As early as 1982, Toys "R" Us marketing aimed to embed its brand in children's identity. That's when they debuted commercials featuring adorable kids singing what has been called the most "the most iconic and lasting jingle in retail history"[10] and one that "lodged itself in the brains of a generation of kids."[11] I was a parent, not a child, in the late 1980s, and more than thirty years later—as I write this—I still can't get it out of my head. The opening lines are, "I don't wanna grow up, I'm a Toys 'R' Us kid." It's impressive that the advertising agency that spawned the jingle managed to cram "Toys 'R' Us kid" five times into a mere eight lines. It all looked like so much fun that even now it's hard to step back and criticize the lyrics. But we do need to question the ethics of persuading kids to bond with a toy store whose sole interest in them was financial.

News coverage and analysis of the bankruptcy focused mostly on extolling its first fifty or so years of incredible success its days as the top toy seller in the world and speculating about why it failed (with blame placed on children's screen time, Walmart, Amazon, and Target, private equity companies, and—bizarrely—millennial parents).[12] The company was enshrined as an "American institution," on par with McDonald's.[13]

Notably missing was a discussion of how and whether children were affected by its incredible success and ultimate failure. From the perspective of what children need, or what's best for them, it's hard to see how children benefit from a chain of giant toy stores that spend millions of dollars to hook them on the thrill of shopping for what were mostly junk toys—any more than they benefit from a fast-food restaurant whose profits depend on kids consuming mostly junk food.

The company didn't market particular toys to children. They left that dubious honor to toy companies, whose advertising was often required to mention that their products were sold at Toys "R" Us.[14] The chain did carry some playthings proven to benefit kids—art supplies and puzzles, for instance, blocks and other kinds of free-form building sets. But the toys dominating the store were the ones most advertised. These tended to glorify violence (action figures from violent PG-13 movies, for instance) and sexualization (such as Barbies, Bratz, and Monster High fashion dolls) and to undermine creative play (any of the myriad toys linked to media icons, which arrive with already formed personalities, and chip-enhanced toys that talk, beep, and move on their own, limit children's involvement to button pushing). From the perspective of Toys "R" Us, what customers bought didn't matter as long as they bought something. The company marketed longing and desire—aiming to instill in children a belief that buying something—anything—at Toys "R" Us was key to their happiness.

Toys "R" Us may have been an American institution, but it's one that—like McDonald's—is emblematic of the American propensity for overconsumption. Their masterful marketing was designed to get children into Toys "R" Us stores brimming with an inventory of up to thirty thousand toys optimally displayed to entrance kids in hopes that they could nag their parents into impulse buys. The company was so good at this that a cartoon appeared in a Florida newspaper of a couple emerging from a Toys "R" Us store laden with packages and captioned "Broke 'R' Us."[15]

Some Toys "R" Us locations even ran free camps for children at its stores, three afternoons a week for about six weeks each summer. These camps, set amid displays of toys for sale, included activities such as craft projects, bubble blowing—and a tour of the store. For parents of young children looking for ways to keep their kids occupied during long summer days, especially those who couldn't afford the expense of day care or actual camp fees, the idea of a free camp may have seemed like a great deal. Except they weren't really free. Expecting young children to spend two hours a day in a toy store without nagging their parents to buy something is as absurd as expecting alcoholics locked in a liquor store to refrain from drinking, or gambling addicts to be in a casino without playing the slots. One mother of little girls aged two and five, who attended the camp regularly for its six-week duration, told the *Washington Post* that "there's only about two times I've left without buying."[16]

As I wrote in *Consuming Kids*, one of the best definitions of commercialism I've seen is offered by James Twitchell, who has written extensively about advertising and commercial culture, though not, I may add, from a particularly critical point of view. According to Twitchell, commercialism consists of commodification and marketing. The former is characterized by "stripping an object of all other values except its value for sale to someone else." The latter involves "inserting that object into a network of exchanges, only some of which involve money."[17]

Twitchell was talking about objects as the things sold to us. But in the marketing world, children are also viewed as things to own, perhaps for life. The objectification of children was exemplified in the 1990s by a quote from Mike Searle, president of Kids "R" Us: "If you own this child at an early age, you can own this child for years to come. . . . Companies are saying, 'Hey, I want to own the kids younger and younger and younger.'"[18]

At PlayCon, I am reminded that the digital marketers replacing Toys "R" Us in the kids market also want to "own" children—and that their tools for doing so are even more effective.

And now I am listening to Armida Ascano, chief insights officer at the market research firm Trend Hunter, give a twenty-minute talk titled "Generation Z: What Fires Up the Youth Generation." According to Pew Charitable Trusts, Generation Z (Pew calls them post-millennials) begins around 1997.[19] Despite her talk's title, Ascano announces that she doesn't think of Generation Z as a generation—instead, the cohort is made up of tribes. She lets that sink in for a beat.

Tribes are traditionally described as social groups "made up of many families, clans, or generations that share the same language, customs, and beliefs."[20] Tribes are characterized by their members' deep emotional, familial, and/or spiritual bonds with each other and their leader, and by the beliefs and values that bind the group together. Tribes also can have positive and negative ramifications. On the one hand, they can provide members with a deep sense of security and belonging. On the other hand, they can be politically and socially divisive and parochial.

When I attended PlayCon, tribes and tribalism were attracting a great deal of attention from U.S. pundits decrying a deeply divided country. That same year, in her book *Political Tribes*, Yale Law professor Amy Chua commented, "The Left believes that right-wing tribalism—bigotry, racism—is tearing the country apart.

The Right believes that left-wing tribalism—identity politics, political correctness—is tearing the country apart. Both are right."[21]

After years of attending advertising conferences and reading trade publications focused on how to market to children, I've concluded that the central question for marketers is not whether a trend is good or bad for kids, or anyone else, but whether, and how much, a trend or societal phenomenon can be monetized. Several important discussions can be had about the impact of tribes on modern society. But that's not why Ascano brings them up. Instead, she tells us that tribes can be organized around goals or causes *but* (my italics) in "the best-case scenario for those of us in the room," members of Generation Z will affiliate with tribes centered around brands.[22]

Today, according to the *London Times*, "teen tribes" can be identified by their carefully curated appearance. But this isn't new. Material goods—in particular, clothing—have long been used to signal affiliation. In the 1920s, flappers wore chemise dresses. In the 1940s and early 1950s, girls who loved pop music wore circle skirts and short white socks and were called "bobby-soxers." In my 1960s high school class, girls in sororities (or sorority girl wannabes) wore a single gold circle pin attached to the Peter Pan collars of their blouses. Rebels with and without a cause wore tight jeans and white T-shirts.

Among the tribes identified in the *London Times* article is one labeled "e-people" (the "e" stands for "electronic").[23] What binds e-people together is their "love of video games, anime and spending copious amounts of time online." They tend to wear "hoodies, baggy clothes, short-sleeved tops over long-sleeved rollnecks and chains on their jeans. E-girls wear their hair in pigtails, with two thin strands left out at the front, often dyed a bright colour. Make-up-wise, it's bold eyeliner, dark lipstick and small doodles (often hearts) beneath their eyes. E-boys have longer hair with curtain fringes and chipped black nail polish."

These groups use(d) clothing and accessories to identify

themselves to other people—to separate insiders from outsiders. But their material expressions of affiliation are mere vehicles to signal belonging to a particular group with a particular set of values, rituals, and mores. In tribes devoted to brands, the product is a de facto leader, and its members are vehicles for generating profit. In addition to being brand loyal customers, they also function, in this digital age of "shares," "likes," and "comments," as a vast and unpaid sales force.

Marketers have long worked to persuade customers to encourage friends to buy their products. Before social networks, marketers handed out free samples of products like CDs, for instance, to kids identified by other kids as "cool." Word-of-mouth, or peer-to-peer, marketing has always been an important tool for advertisers. Digital social networks enable word-of-mouth marketing on steroids. McDonald's, for example, has millions of followers on Facebook and Instagram. Each time someone "likes" a McDonald's post, everyone in their network gets what is essentially a mini-advertisement for the fast-food giant.

I'm sure the predigital purveyors of clothing and accessories would have loved a chance to make their brands more central to bonds formed by various groups—but they didn't have the tools to do that. Now they do, thanks to the internet and, in particular, to social media. They hone their messages by gathering and analyzing personal information about their customers and potential customers. They catalyze online interactions with customers and encourage them to connect with each other.

According to Vivid Brand, a British marketing company, brand tribes are "groups of people emotionally connected by similar consumption values and usage, who use the social 'linking value' of products and service to create a community and express identity."[24] The primary emotional connection among brand-oriented members of a tribe is their mutual devotion to a product, or group of products, from which they derive their identity.

Marketing experts advise companies to "discover the shared

characteristics that define your tribe, speak to the changes and challenges that its members are experiencing, and create folklore and stories that will strengthen the bonds of your tribe and stoke its passion for your brand. In turn, your tribe members will help you socialize your messaging, evangelize your products, and amplify your brand."[25] In plain language—in the age of social networks—devotees of a brand tribe are valued because they provide companies with tons of free marketing. Brands bent on cultivating tribes should gather personal information from members to provide curated content that speaks to members' vulnerabilities and keeps them emotionally engaged.

That companies exploit our children's deeply human need for affiliation is worrisome. It's like the way we might worry about anyone developing an emotional bond, and identification, with a group centered around a leader whose primary interest is not in the members' wellbeing but in using them to generate profit, or for any personal gain.

The imperative to create brand tribes is just one component of Ascano's message. She tells us that companies need to identify the emotional vulnerabilities of their potential customers and to persuade them that buying into a brand will assuage these needs. She tells us that Gen Z's need for affiliation is an expression of their search for identity—and for Gen Z, identity is fluid. They have a lot of questions, she says, that society doesn't have answers to. Brands "should" support Generation Z's healthy growth and development by tackling the "tough topics." She shows us her favorite advertising campaign from 2017. It's from Axe, a line of male grooming products owned by Unilever. The commercial is designed to tug your heartstrings and to position Axe as "woke" about race and gender. It features a diverse group of attractive young men asking questions like "Can I wear pink?" and "Can I experiment with another guy?"

It's an ironic twist on Axe's previous campaigns—which have

a history of blatantly sexualizing and objectifying women.[26] Now Axe is positioning itself as a force for good. The underlying message is that society is failing boys and that brands like Axe—not parents, advocacy groups, religious and educational institutions, or civic and social organizations—can provide them with answers to tough questions and have their best interests at heart.

As Ascano celebrates brand tribes, and this particular Axe campaign, I find myself thinking that her whole presentation is emblematic of a deeper and more pervasive problem. In her discussion of tribes, she conflates the terms "cause" and "brand." Yet they do not mean the same thing. The former is a movement based on an ideal, or unifying principle. The latter is a marketing term evoking a set of characteristics that distinguishes one commodity from another similar one. The myth that there is no difference between entities whose primary goal is to generate profits and entities whose primary goal is to promote the common good provides fertile ground for nurturing greed in a society.

In the 1970s, the psychologist Urie Bronfenbrenner expanded our understanding of the realms of influences on children to include more seemingly remote societal forces—the economic, cultural, religious, and political systems in which they live.[27] The most obvious examples are dreadful. The 2008 recession, which raised unemployment, added two million children to the thirteen million already living in poverty in the United States.[28] The state of Michigan's decision to save money by switching the town of Flint's water source caused many kids to be exposed to lead.[29] Violent political or religious conflicts around the world leave children homeless, orphaned, or worse.

Not all societal events and changes affecting children are dramatic enough to capture headlines. Some, like the digital revolution, sneak up in the guise of mere entertainment, or convenience, or that double-edged sword "progress." And some, like the escalation of commercialism in the lives of children, are the result of overlapping phenomena—inadequate regulation of corporations and advances in digital technology that enable the lucrative ubiquity of tablets, smartphones, artificial intelligence, and wearable tech.

Today, the language, values, and strategies of marketing permeate our civic, spiritual, and personal lives. Perhaps the most obvious example is the whole concept of branding. Branding gets conflated with reputation—but their meanings are different, and the differences are important. Branding is marketing jargon for how companies choose to represent themselves to consumers and to differentiate themselves from their competitors. It's based on a self-directed image, and may or may not accurately encompass the reality of what the entity being branded really does. A reputation is based on the actions of a person or organization and how these actions are perceived by people familiar with them. As one marketer said, "a good brand impression can be created through a smart advertising campaign, but you can't advertise your way to a good reputation."[30]

The process and techniques of branding focus solely on image. The goal of branding is to create an emotional tie with potential customers that morphs into brand loyalty, the Holy Grail for marketers.[31] Brand loyalty is one of the primary reasons that companies market to children. It can last a lifetime and brand-loyal customers are not easily dissuaded from their brand of choice.[32] James McNeal, a psychologist who has written extensively about how and why companies should market to children, estimated in 2000 that a lifetime brand-loyal customer could be worth $100,000 to an individual retailer.[33] (In 2022, with inflation, lifetime, or cradle-to-grave brand loyalty, was worth $164,760.)

Blind taste tests show that people prefer better quality products when they don't know the brand of the products they are trying—but not when they do know the brands. Branding is so powerful that it even displaces our senses. Preschool children identified food wrapped in McDonald's wrappers as tasting better than the same food wrapped in unbranded packages—even though all the food they tasted was from the same source.[34] The same is true for food branded with licensed characters, like Shrek, Scooby-Doo, or Spiderman.[35] Brand-loyal customers are likely to stop shopping around. They're likely never to switch to a better-quality product and not to notice changes, such as price fluctuations, in their preferred brand.[36]

As brand loyalty increases, customers are less sensitive to changes in how much that brand costs. They are also less likely to notice or be susceptible to competitive promotions sponsored by other companies.[37] Brand loyalty allows companies to raise prices without much complaint and to save money on marketing. Brand loyalty means that people might keep buying a brand even if their original reasons for purchasing it—such as cost—may no longer be valid and even if it is in their best interest to buy the same kind of product from a different company. Brand loyalty is beneficial to a company but not necessarily good for a customer.

When in January 2016, Donald Trump, who was then the Republican nominee for president of the United States, famously said, "I could shoot people on the streets and not lose support,"[38] the convergence of marketing and politics reached a new and troubling apex. Lots of people, regardless of their political affiliation, were appalled. But Trump, speaking simultaneously as a consummate marketer and brand, was merely applying the conventional wisdom about brand-loyal customers: they will stick with a brand regardless of changes in price or quality. The stories companies tell us about themselves and their products create the foundation of branding, and these stories are ubiquitous. In today's digitized world, we don't need to seek out marketing messages, they find us. What takes effort is seeking out facts about corporate behavior.

Apple, for instance, revered among marketers for its devotedly loyal customers and for the emotional connections it generates with consumers, has topped *Forbes*' list of the world's most valuable brands every year since the list began in 2010.[39] In January 1984, Apple launched its now iconic TV commercial during the Super Bowl.[40] It features a dystopian, dictatorial state, reminiscent of George Orwell's book *1984*, that is disrupted and presumably destroyed by a young woman—who is the only spot of color in this dismal, gray world—hurling a hammer at an image on a screen. The screen explodes into a bright white light, and we see and hear the following words: "On January 24th Apple Computer will launch Macintosh, and you'll see why 1984 won't be like 1984." The ad launched a decades-long marketing trajectory that positioned Apple as antiauthoritarian, pro-freedom, pro-democracy, and pro-nonconformity.

In 2015, Apple's CEO Tim Cook reinforced this branding when he was interviewed about privacy on National Public Radio. He described privacy as "a fundamental human right,"[41] and claimed that Apple was more protective of its customers than other tech companies were. Yet in 2017, Apple began allowing the highly authoritarian, nondemocratic Chinese government to access data from all Chinese citizens who stored data in Apple's iCloud—including their contacts, photos, text files, and calendars.[42] Four years later, the *New York Times* quoted Nicholas Bequelin, East Asia director for the human rights group Amnesty International, describing Apple as a "cog in the censorship machine that presents a government-controlled version of the internet." Bequelin went on to say that "if you look at the behavior of the Chinese government, you don't see any resistance from Apple—no history of standing up for the principles that Apple claims to be so attached to."[43]

Apple is just one example of a corporation whose actions belie its branding. Given that the primary goal of most corporations is to make money for their stockholders—not to promote democracy or wellbeing—corporate hypocrisy, though upsetting, is not unexpected. In fact, knowing that what's best for a corporation—profits—can easily be at odds with what's best for humanity, we ought to be able to rely on civil society to keep corporate behavior in check. Instead, the line between the commercial world and the civil and spiritual institutions whose stated mission is to ensure and even elevate the wellbeing of people and the planet is increasingly blurred. Beginning in the 1980s, the U.S. government significantly reduced its regulation of corporations and ceded oversight of corporate behavior to the corporations themselves. Today, while corporations have many of the rights and privileges of individual citizens, we the people have adopted corporate behaviors and values, including branding, as normal, or even laudatory.

When commercial values dominate children's environment, kids are in danger of losing out on exposure to some of the best human values, such as altruism, generosity, nonconformity, and critical thinking. Branding, the creation and marketing of image, is no longer relegated to commerce—it's part of daily life. In such a society, even when children are not targeted directly by marketers, they absorb the cynical message that everything is for sale, including government, religion, education—and themselves. I've come to think of the phenomenon as "trickle-down branding."

Presidents of the United States, for instance, have been working with marketers and advertising agencies for over fifty years. From Lyndon Johnson through Ronald Reagan, both George Bushes, and Bill Clinton, presidential candidates upended the tradition of campaign advertising that at least metaphorically addressed issue and policy differences between candidates by focusing instead on evoking emotion.[44]

It was in 2008, however, that the demarcations between U.S. presidents, brands, and marketers got really blurry. That's when *Advertising Age* named presidential candidate Barack Obama Marketer of the Year.[45] Keith Reinhard, chairmen emeritus of the global marketing agency DDB Worldwide, noted, "Obama is three things you want in a brand—new, different and attractive. That's as good as it gets."[46]

Two years later, in 2010, Staci Zavattaro, then a doctoral student in public administration at Florida Atlantic University, published a prescient article in the academic journal *Administrative Theory and Praxis* that detailed three reasons why presidential branding is worrisome. First, branding turns the president into a commodity that can be bought and sold. Next, political marketing will shift from being "platform-centered to candidate-centered." And last, "As a productized brand, the president becomes a simulation for a leader . . . presidents will feed into the simulation of an ideal image rather than an ideal leader."[47]

Presidential branding reached yet another level in the 2016 election of Donald Trump, whose brand-marketing acumen is his undisputed accomplishment, as president of the United States. His blatant and unapologetic use of public office to promote his private brand was unprecedented.[48] Back in 2010, Zavattaro didn't anticipate a president who was already synonymous with his commercial brand, but, with that twist, all three of her predictions came true.

To her first point: politicians, foreign government officials, and lobbyists certainly behaved as though the president could be bought. According to Citizens for Responsibility and Ethics in Washington (CREW), during Trump's first year in office, more than thirty members of Congress stayed at Trump-owned properties, and forty special interest groups held events there. Six foreign government officials made appearances at his properties, and eleven foreign governments paid Trump-owned businesses.[49]

To Zavattaro's second point: Trump's celebrity did more than overshadow his party's platform—it obliterated it. In the 2020 election, Republicans didn't even bother to create a platform.[50] And finally, to her third point: During his presidency, Trump was well known for his lack of interest in policy detail, his unwillingness to prepare for important meetings, and his lack of interest in the details of governing.[51] But his love for the pomp and circumstance of the presidency is well documented.[52]

One could argue that Trump was merely a one-off example, since in 2020 a president was elected who, unlike Trump, is similar to his predecessors, who used marketing as a tool to secure and promote their presidencies rather than using the presidency as a tool to promote their marketing. But it's certainly possible that the barriers Trump smashed between politics and self-branding set precedents that will make it easier for others to follow suit.

Meanwhile, the churches, temples, mosques, or synagogues children and their families might attend are also branding themselves. Marketers, or "social communicators," sell their services to religious institutions with articles like "Ten Common Branding Mistakes that Churches Make," which begins: "Effective branding for churches has never been more important. We live in a media-driven age where first impressions last and a church's brand presence can literally be the difference between someone coming to visit a church or not."[53] When houses of worship brand themselves in the same manner sellers of hamburgers do, those institutions become commodities to buy and sell, too.

Equally worrisome is that teachers are becoming official brand promoters. In 2017, a story appeared in the *New York Times* with the headline "Silicon Valley Courts Brand-Name Teachers, Raising Ethics Issues."[54] Edtech companies select teachers who have a social media presence or speak at education conferences to act as company spokespeople. The teachers-cum-marketers receive all sorts of free stuff, including ed tech products to use in their classrooms, in exchange for promoting that company's product or brand to their followers on social media or in their teacher-training workshops. As one teacher said about her relationship with an edtech startup: "I will embed [the company] every day in my brand." The teachers involved don't seem to see a conflict of interest, but it clearly may affect their work in the classroom, much like research in medicine shows that gifts from drug companies can influence the ways health practitioners treat the people they serve.[55]

Today, as culture is increasingly commercialized, many of children's traditional real-life heroes—athletes, musicians, television actors, and beloved fantasy characters—are all selling something. Over the past decade, a plethora of books with titles like *You Are the Brand*, *You Are a Brand!*, *You Are Your Brand*, and *U R the Brand* sell the notion that each of us needs to be packaged and marketed. I'm not suggesting that kids are reading these books, but the message is trickling down to them, especially through social media influencers. The term "influencer," according to *Wired*, is "inextricably tied to consumerism and the rise of technology." It's "shorthand for someone (or something) with the power to affect the buying habits or quantifiable actions of others by uploading some form of original—often sponsored—content to social media platforms."[56] Among kids aged eight to twelve, three times as many want to be influencers as want to be astronauts.[57]

Influencers can be celebrities posting paid shout-outs to brands on Instagram. But they can also be seemingly regular people, including kids, who rack up enough followers to make potential product endorsements valuable to those who make and market consumer goods. Over time, these glorified salespeople, themselves, become brands.

Successful social media influencers can make big bucks from companies paying them to push products. In 2020, nine-year-old Ryan Kaji earned $29 million on YouTube, making him the highest paid YouTube star that year. His channel Ryan's ToysReview (now called Ryan's World) had 2.2 billion followers. His brand Ryan's World licenses more than five thousand products,[58] including toys, housewares, and clothing,[59] which grossed more than $250 million in 2021.[60]

But whether we generate income from our posts or not (and most of us don't), social media encourages everyone, including children, to self-brand. Social media prioritizes image and is not so much about truth, at least not the whole truth. We curate our posts, and what we opt to share of ourselves may or may not reflect what's really going on in our uncurated lives. Platforms popular with teens and preteens, like Instagram, TikTok, and YouTube, teach kids to sell themselves—if not for money, then for virtual approval in the form of "likes," "shares," "friends," and "followers."

As with so many of digital technology's innovations, it's possible for social media to play a positive role in children's lives.

Producing videos for TikTok and YouTube can be wonderfully creative. Facebook and Instagram can facilitate connections for kids who don't fit easily into their communities' norms. The problem is that social media sites are home to some of the worst tech industry business practices. TikTok's infinite scroll is designed to keep users on the site indefinitely. Snapchat's streaks, in which two people respond to each other's posts for as long as possible, are designed to keep kids constantly returning to the platform. Instagram's photo editing feature allows users to correct perceived flaws in their appearance, including the ability to slim waists and legs, sending the clear message that our genuine appearance isn't good enough for sharing. Viewing these altered photos seems to negatively affect the body image of teenage girls, who already have tendencies to compare themselves with others.[61] Girls, who have long struggled with comparing themselves with models and other celebrities, now have to cope with falsely glamorized shots of their peers.

Social media platforms promote engagement by tagging posts and pages with metrics such as likes, shares, and numbers of views and friends, subjecting kids to extremely public, and possibly daily or even hourly, popularity contests. Depending on the numbers, these metrics can be elevating or shaming. I spoke to a fifteen-year-old who told me about friends who deleted Instagram posts that didn't have a couple of hundred likes. Heavy social media use is linked to unhappiness in adolescents and dissatisfaction with their lives. Interestingly, while eighth and tenth graders who use social media two hours a week are happier than kids who don't use it at all, the survey also found that those kids who are extremely light social media users are happier than kids who use social media two hours a day, which is not even considered heavy use.[62]

While social media platforms like Instagram and YouTube claim that they don't allow users under thirteen years of age, millions of younger children lie about their ages to create accounts.[63] The time preteen girls who pretend to be teenagers spend with social media is linked to increased dissatisfaction with their appearance.[64] Yet, in March 2021, an internal memo leaked to BuzzFeed revealed that Instagram, which is owned by Meta, was creating a children's version of its platform.[65] Facebook was already directly targeting young children through Messenger Kids, an iteration of Facebook's instant messaging service, which launched in 2017.[66] Two years later, it was discovered that despite the company's claim that children using Messenger Kids could talk only to users approved by their parents, children in group chats were in unsupervised contact with strangers.[67] Facebook fixed what it called a "design flaw," but critics saw it as emblematic of the company's inability, or unwillingness, to truly protect children.[68]

At a congressional hearing just three days after news of Meta's plan for a kids Instagram surfaced, Mark Zuckerberg, Meta's founder, CEO, and principal shareholder, tried to allay concerns about children using social media by saying that "the research that we've seen is that using social apps to connect with other people can have positive mental-health benefits."[69] What's especially egregious about Zuckerberg's testimony and Meta's determination to foist Instagram on preteens and younger kids is that the company's executives have known for years that the popular social media site is toxic for a significant percentage of teenage girls.

In September 2021, the *Wall Street Journal* obtained internal memos documenting that, according to Meta's own researchers, "thirty-two percent of teen girls said that when they felt bad about their bodies, Instagram made them feel worse" and "comparisons on Instagram can change how young women view and describe themselves."[70] Another presentation by the company's researchers even showed that "among teens who reported suicidal thoughts, 13 percent of British users and 6 percent of American users traced the desire to kill themselves to Instagram."[71]

These aren't the only leaked documents indicting Meta's executives for their willingness to harm teenagers. In 2017, an internal document surfaced in Australia describing how Facebook can help advertisers target teens based on their emotional states, including when they feel "'insecure,' 'defeated,' 'anxious,' 'silly,' 'useless,' 'stupid,' 'overwhelmed,' 'stressed,' and 'a failure.'"[72]

And that's not all. This year, an investigative report from the advocacy group Reset Australia found that Facebook allows targeted advertising to teens under eighteen whose profiles suggest concerning predilections such as gambling, drinking, vaping, "extreme weight loss," and online dating.[73] While the ads Facebook approved were not explicitly for alcohol, unhealthy eating, or online dating sites, they certainly encouraged these interests. For instance, one included a cocktail recipe, another asked if girls were "summer ready," and a third proclaimed "find your gentleman now."[74] The cost for such advertising? It's pretty cheap. According to the report, "Reaching a thousand young people profiled as interested in alcohol will cost advertisers around $3.03, $38.46 for those interested in extreme weight loss, or $127.88 to those profiled as interested in smoking."[75]

In September 2021, the aforementioned leaks and intense pressure from advocacy groups,[76] forty-four state attorneys general,[77] and members of Congress[78] led Adam Mosseri, who heads Instagram, to announce that Meta was putting its kids version of Instagram on "pause."[79] In a blog post, Mosseri said, "We will work with parents, experts, policymakers, and regulators, to listen to their concerns, and to demonstrate the value and importance of this project for younger teens online today."[80] Really? Given Meta's dismal track record with children, isn't the only value and importance of Instagram Youth to train younger kids to use the actual site when they turn thirteen?

Another leaked memo answered its own question "Why do we care about tweens?" with the following: "They are valuable and untapped audience." And went on to say, "Our goal with MK [Messenger Kids] is to gain messaging primacy with tweens . . . which may also lead winning with teens.[81] "At one point, the company contemplated leveraging children's playdates to "drive word of hand/growth among kids."[82]

Facebook's assistance to advertisers whose strategy is to target teenagers according to their vulnerabilities to potentially dangerous behaviors and to exploit their interests is, of course, appalling. It's also a cautionary tale for how Instagram will treat (or mistreat) younger children. Even without exposing kids to unscrupulous advertising, social networks are inherently problematic for them. Encouraging children to self-brand on social media turns childhood itself into a marketplace where authenticity is devalued and appearance—brand image—is what matters. And given that so much of the world around them has also embraced branding, what's left to counterbalance that perception? Are these really the values we want to pass on to our children?

4

Browse! Click! Buy! Repeat!

When you reduce friction, make something easy, people will want to do more of it.

—Jeff Bezos, CEO, Amazon

Last year at an industry conference on marketing to kids, I listened to a tech executive advise attendees to "reduce friction." Since the friction-reducing I'm familiar with involves either WD-40 or family therapy, I was momentarily bewildered. Then I got it. When marketers talk about *reducing friction*, they mean removing any barriers or hurdles between the products companies offer and the consumers who purchase them. Amazon's innovations like one-click buying and increasingly short delivery times brought Jeff Bezos's goal of frictionless commerce to the forefront of e-commerce today. For children immersed in the digital world, reducing friction manifests as in-app advertising for purchases that promise to make whatever games kids play so much more fun. Another example is YouTube's default setting for autoplay,* which entices all of us, including the millions of children who use the platform, to watch

---

* Autoplay is turned off in the default settings for YouTube Kids, but research suggests that many more children watch YouTube than watch YouTube Kids. See Jenny S. Radesky et al., "Young Children's Use of Smartphones and Tablets," *Pediatrics* 146, no. 1 (July 1, 2020), doi.org/10.1542/peds.2019-3518.

one advertising-laden video after another without even having to exert the minimal effort to click, swipe, or tap.

While the idea of frictionless commerce may have originated in the tech world, brick-and-mortar businesses that target kids also aim to reduce friction, often by seamlessly integrating things to do with things to buy.

Conversations I've had with parents about raising kids in a hyper-commercialized world have gone like this. "I feel like children's play spaces are designed to funnel us into an inescapable shopping opportunity," complained a friend with two young children. "When we visited my brother in Ohio last year, they were excited to visit Entertainment Junction. At first it was great. You walk through a long tunnel of model trains and things to climb on—perfect for active kids like mine. But then we were dumped directly into this huge store brimming with things my kids love. Honestly, I'm okay with refusing to fulfill my children's request for stuff. But it's hard on all of us when they're continually confronted with things to buy even when we're not at a store."

A six-year-old's father said, "My daughter loves gymnastics, so I signed her up for this kids' gym. I didn't even think about the store right in the lobby. You can't avoid it. Sure, it sells leotards and things you might need for gymnastics—but it also sells dolls and sparkly headbands. I have to tell her no every time we go to her class." As another parent put it, "Especially in winter, when we're cooped up because it's too cold and miserable to spend a lot of time outside, it feels like every place we go devolves into a decision about whether or not to buy something."

And this from yet another parent, "My kids are starting to believe that every experience should end in acquiring stuff. And *getting* the stuff seems to be more important than actually having it. Once the thrill is gone, and it often goes pretty quickly, they move on to wanting the next thing."

This last observation gets to the crux of one of the most serious consequences for kids growing up in a heavily commercialized culture. They are repeatedly exposed to the seductive and destructive materialistic messages embedded in all corporate marketing, most prominently that the things we buy will make us happy. But they don't, at least not in any kind of sustained way.[1] The belief that happiness resides in our next purchase combined with the fact that it doesn't is a boon to the makers and marketers of the things we buy.

Here's how it works. If you believe that things will make you happy, what do you do? You buy a thing. But since things don't make us happy, buying the thing doesn't make you happy—at least not for very long. But if you still believe that things will make you happy, what do you do? You buy another thing. A different thing. A bigger thing. A better thing. And so on.

Our deep attachment to objects, and our tendency to invest them with meaning and power beyond their physicality, is an ancient trait dating to long before the invention of modern advertising and marketing. Objects have long served more than utilitarian needs and held symbolic meaning central to religious rites and spiritual practices. This impulse to invest objects with deeply personal meaning begins in early childhood.[2] Perhaps because young children haven't yet acquired the capacity for abstract thinking, they need to depend on what is tangible more than adults do.

As babies transition from total dependence on their caregivers to their first steps toward independence as toddlers, they might latch on to something, perhaps a blanket or a stuffed toy, that comes to embody love, safety, security, power, companionship, and more. I knew a little boy, whom I'll call Alex, who became devoted to an old couch cushion, which he called "Cushy." He derived great joy and comfort from Cushy and refused to be separated from it. Unfortunately, one day Cushy—being an elderly cushion—imploded. Alex was distraught until his mom removed the remaining stuffing and sewed the cushion's cover into a flat square, replete with button eyes and a yarn tail. This new Cushy iteration continued to be a source of comfort and security over the next few years.

Gradually, without even realizing it, Alex began to take on the attributes he once assigned to Cushy, and he came to recognize Cushy for what it was—a piece of cloth with no special powers. Once that happened, Cushy stopped being centrally important to Alex's sense of security and wellbeing.

Like most babies, Alex was born with the potential to value objects and imbue them with great importance. He was also born with the potential for other important attributes: creativity, a deep connection to the natural world, a capacity for spiritual experience, and a deep love and need for other people. Which of these develop and which do not depends in large part on the values and experiences Alex encounters as he grows. These are shaped by his family, his community, and also by larger societal forces, including the economic system in which he lives.

Twenty years ago, as marketing to children escalated, I attended a gathering of professionals working in children's media—many of whom produced well-respected programs on both public and cable television. After listening to the conversation for a while, I commented that rather than talking about how programming could improve children's lives, the conversation was all about money. There was silence, and then one of the television executives spoke up. "Capitalism won," he said. He was being ironic—referring to the relatively recent collapse of Soviet-style socialism—but the message was clear. Public funding for children's programming was diminishing and was unlikely to increase. Children's shows had to be profitable, and that meant generating income through advertising and/or by licensing characters to sell food, clothes, toys, and accessories to kids.

Today, when parents describe their quandaries about children and commercialism, I find myself thinking about that television executive's tossed-off comment about capitalism. Children in the United States today are growing up in a culture profoundly shaped by a troubling combination of what psychologist and philosopher Shoshana Zuboff and others have called "surveillance capitalism," which is fueled by the tech industry's mining of personal information for profit, and "corporate capitalism," which is dominated by huge privately owned businesses whose primary obligation is to make money for their shareholders. Today, these forces combine to create what's best described as "consumer capitalism," a sociopolitical economic system driven by, and in thrall to, consumption.

Consumer capitalism depends on a population that must be primed continually to buy the things that businesses sell. Corporations, mass-producing goods for sale, stoke consumer demand through mass marketing designed to foment desire for products by blurring the distinction between wants and needs. This century's stunning advances in digital technology, nurtured in a political climate that has been decidedly anti-regulation, provide more numerous, more sophisticated, and more subtle avenues for companies not just to market their wares widely but to target their advertising precisely to exploit and even mold individual needs, wants, and desires.

So, what does this have to do with children? Any economic or social system depends on a population made ready and willing to adopt the values, behaviors, and attributes that perpetuate that system.[3] When corporate executives talk about reducing friction, some of what they mean has to do with reducing external barriers to buying, but it also means reducing or eliminating our intrapsychic friction—the cognitive and emotional brakes that enable us to set limits on consumption. For that reason, kids are not just fair game for advertisers—they are essential targets. Their immature capacities for judgment and impulse control render them especially susceptible to marketing messages.

As I describe in chapter 1, the very structure of a child's brain is shaped by early experience,[4] laying a foundation for future behaviors, attributes, and values—including the ingrained expectation, described earlier in this chapter, that purchasing stuff is routinely part of any excursion. The parent lamenting his children's expectation that every experience outside of the house automatically includes buying something is describing kids who have been primed to consume. The purpose of an individual ad might be to sell a particular product. In aggregate, however, advertising sells its targets on the virtues and benefits of unquestioning and ultimately insatiable consumption.

I've long believed that marketing to children in and of itself is harmful, even apart from the products being marketed. My reasoning was based on a slew of research showing that advertising is a factor in so many of the problems facing children today: childhood obesity, eating disorders, precocious sexuality, youth violence, the erosion of creative play, and more.* I saw the acquisition of materialistic values as one of many problems linked to advertising—on par with, but not greater than, childhood obesity, precocious sexuality, youth violence, and others.

For many years, when I was asked to name the worst problem linked to marketing to kids, I often explained that the harms caused by exposure to marketing depended on children's individual predilections and vulnerabilities. So, if your child is vulnerable to overeating and poor nutrition habits, then marketing junk food seems like the worst. It's also linked to poor body image—so if your child is vulnerable to insecurities about physical appearance, or eating disorders, then that's what gets your attention. Marketing is also linked to aggressive behavior, so if your child is susceptible to violent messages, then you might see that as the worst potential effect of marketing to children. And so on for all marketing-related problems, including decreased creativity, increased family stress, and the acquisition of materialistic values. I believed that the reason we should work toward ending all marketing to children was because the harms were so broad and far reaching that it made sense for advocates to work together. I still stand by that belief. These days, however, corporate marketing's incessant celebration

---

*See my books *Consuming Kids* and *The Case for Make Believe*.

of consumption and the materialistic values that drive it loom larger in my concerns.

Based on a variety of dictionary definitions, as well as on the work of researchers around the world, I define "materialistic" as describing people who tend to value wealth and possessions relatively high compared with how they value spiritual, intellectual, cultural, or relational aspects of being human.[5] I do, however, have to admit to a particular fondness for the definition someone named Cheyenne posted in 2005 on urbandictionary.com: "The act of caring more about THINGS than people; judging yourself and others on the cost of your stupid things."[6]

Mounting evidence suggests, at the very least, that advertising exacerbates children's materialism. Several studies demonstrate that kids with more exposure to advertising are more materialistic than their peers and that excessive materialism and the unchecked mass consumption it drives are harmful on multiple fronts.[7] One increasingly obvious and obviously dire consequence is devastation to the planet—the depletion of natural resources and the increase in pollution and global warming.[8] As a team of scientists wrote in the medical journal *Lancet*, "Climate change is a direct consequence of development based on fossil fuel that started in the European industrial revolution. An accurate simplification is to say consumers, rather than people, cause climate change."[9] Since much has been written about the links between consumption and environmental degradation, I won't devote a lot of space to it here. I do, however, want to discuss a huge and environmentally lunatic trend in toy marketing: plastic collectibles for kids.

"Collectibles" is toy-speak for brands whose profits rest in convincing kids that one toy won't satisfy them, instead they have to "collect" a whole line. I've seen collectibles marketed as sets of as many as one hundred. They're often cheap, until parents realize—like the old potato chip ad—that kids are being primed to dissatisfaction with owning only one.

Collecting is a time-honored human preoccupation spanning all age groups, including childhood. Children have long collected shells, rocks, stamps, coins, and all sorts of things. Educators point to the benefits of collecting, which include deepening and expanding knowledge (by learning about the things you collect), improving self-esteem (by engendering a sense of accomplishment and competence), encouraging self-expression (by being able to demonstrate concretely your interests to others), and finding community (by bonding with others who share your passion).

Industries have been profiting from the human impulse to collect since the late 1880s, when tobacco companies included photos of celebrities, including baseball players, in packs of cigarettes. As radio, and then television, brought professional sports into our living rooms, the focus switched to kids and chewing gum packs containing baseball cards. By the 1950s, companies began using the cards to market gum instead of vice versa.[10] McDonald's introduced its iconic Happy Meal and Happy Meal toy in 1979. That same year they began offering themed Happy Meal toys from the movie *Star Trek: The Motion Picture.* Subsequent partnerships routinely feature toys tied to popular films, TV series, and video games.[11]*

Beanie Babies, the Ty Corporation's seemingly endless supply of cute little stuffed animals, were all the rage in the 1990s.[12] Nintendo introduced Pokémon, the grandparent of today's collectibles craze, in 1996.[13] Pokémon began as a video game that morphed into—among other things—trading cards, a television

---

* I should note that in 2021 McDonald's announced that it was going to stop making most of its Happy Meal toys out of plastic (Bill Chappell, "McDonald's Is Phasing Out Plastic Toys from Happy Meals in a Push to Be More Green," NPR, September 21, 2021). If that's the case, and depending on what they substitute it with, it's good news. But as long as they keep using toys, regardless of their composition, to sell fast food to kids, they are still (a) promoting consumerism and (b) encouraging children's unhealthy eating habits. The latter is not the focus of this book, but I think it's important to acknowledge.

series, a movie marketed with fast-food giveaway toys, apps, and most recently an augmented reality game. Its original theme song, which features the line "gotta catch 'em all" sixteen times in just over three minutes is an early example of the kind of discontent fostered by companies today. And cardboard baseball cards have been displaced by plastic—lots and lots of plastic.

A report from the World Economic Forum estimates that by 2050 there will be more tons of plastic in the ocean than tons of fish.[14] An article on the *Forbes* website cites studies that found that trashed plastic is already "in the guts of more than 90 percent of the world's sea birds, in the stomachs of more than half of the world's sea turtles and is even choking the life out of whales."[15]

Leading the market in plastic collectible toys are LOL Surprise Dolls, made and marketed by MGA Entertainment. MGA is the company that took on Mattel's Barbie by marketing the sexualized, big-eyed, big-breasted, tiny-waisted Bratz dolls. Like the Bratz, LOL dolls are the brainchild of Isaac Larian, MGA's CEO and founder. According to *The Atlantic*, the dolls were "essentially reverse engineered: The company wanted to cash in on the unboxing and collectibles trends."[16] A CBS news segment reports that after watching lots of unboxing videos, Larian wanted to create a toy designed so that children could record themselves opening it up. According to Larian, "I said, 'Well we gotta do a toy that's a true unboxing so that every kid can unbox it.'"[17]

Packaging is key to LOL's marketing strategy, and it's an environmental nightmare. To give you a sense of the toys, here's a description of one typical LOL Surprise toy: the doll and a bunch of "surprises" come in a sphere wrapped in shiny pink plastic wrap emblazoned with the message that there are over forty others to buy. Unwrap that and you find a message and another layer of plastic with a couple of stickers. Rip that off and you find yet a third layer and a plastic bottle encased in more plastic. Then there's a fourth layer and a pair of pink plastic shoes also wrapped in plastic. Remove that and you find a fifth layer and a plastic-wrapped pink plastic dress. Finally, unwrapping gets you to a hard plastic ball that opens to reveal a little plastic doll which is, of course, wrapped in plastic.

While I was writing this chapter, children around the world were marching to end climate change, in perhaps the largest environmental action ever. Just three weeks earlier, the American toy industry had celebrated MGA Entertainment by naming LOL Surprise dolls the 2019 Toy of the Year.[18]

The packaging of LOL dolls is rendered even more egregious by the second pillar of MGA's marketing strategy—fomenting perpetual discontent and longing by positioning the contents of each individual package as merely one of a series of collectibles and therefore incomplete on its own. The message to children is clear: you can never, ever have enough. And in the unlikely event that you collect an entire series of dolls, there's a whole new series to start on. LOL dolls are probably the most popular example, but in fact plastic collectibles are a growth industry.[19]

One hallmark of marketing collectible toys is enticing kids to keep buying by trumpeting "blind" boxes or bags, which entice kids to buy more because of the possibility they will hit the jackpot and acquire something rare or missing from their collection. I don't use "hitting the jackpot" lightly. Spending money on unknown and unseen collectibles in hopes of scoring a rare one is like training kids to love gambling.[20] The adrenaline rush, the hopes, the disappointment, and the drive to try again are a lot like the emotional experience of feeding coins in a slot machine in hopes of lining up those elusive three cherries.

Isaac Larian eschewed television advertising when he launched LOL dolls, going straight to unboxing influencers. Kids are still watching TV, however. So, I tuned into Nickelodeon one afternoon not long ago to watch some commercials. Just about every toy advertised was for collectibles, including Hasbro's DreamWorks Trolls Hair Huggers and Lost Kitties and Spin Master's Twisty

Petz. All of these can be found featured in unboxing videos as well. Each toy is relatively inexpensive, ranging from $2.99 to $5.99 at big-box stores. But here's the catch: their marketing pushes the whole series. The ads feature not just one toy but several. Take the Lost Kitties ad, which features predictably adorable kids dancing around waving a couple of Kitties and chanting something like:

> Lost Kitties series two
> 100 you can collect!
> Find Lost Kitties
> Who will you find next?

Those of us who care about children's creative play frequently call out toy companies for deceptively marketing toys as encouraging creative play when in fact the toys rob kids of opportunities to create. What's striking about marketing for Lost Kitties, and other collectible brands, is that companies barely bother to feature play as a selling point of the toys. Instead, they blatantly sell consumption by focusing on the importance of collecting (read "buying") more. And it works. A parent I know put her foot down about buying another LOL doll. She happens to be very concerned about social justice. So her seven-year-old, who was desperately trying to convince her to relent, came up with what she thought was a surefire argument. "Mom," she said, "you know how LOL dolls are made by the Chinese? It's really not fair to the Chinese if we don't buy them." While her argument was ingenious, it was not successful. At least not that time.

Global warming and the degradation of our oceans make obvious the harms of hooking kids on wastefully packaged plastic collectibles and other junk toys. But merely trying to shift their desires to acquiring lots of other kinds of toys, even those that are more sustainable, is not the answer. One media studies scholar points out the hypocrisy of Instagram influencers and mommy bloggers whose posts simultaneously tout the joys of minimalist, uncluttered living and feature ads for high-end consumer products, including exquisite and expensive wooden toys.[21]

I admit to being a longtime sucker for beautifully made toys, including wooden ones. But I also recognize that pushing parents to conflate a minimalist, anti-consumerism lifestyle with substituting cheap plastic toys for high-end wooden ones is just another form of marketing—one that exploits and flaunts the gap between those wealthy enough to purchase expensive toys for their kids and those who aren't. And somehow it imbues tastefully spent wealth with a moral superiority while, by implication, denigrating the morals of lower-income parents who purchase cheaper, mass-marketed plastic toys.

Let me be clear. The problem is not with wooden toys or authentically embraced minimalism. The problem is that even movements eschewing consumerism can be so co-opted by commercialism that they end up encouraging, not discouraging, materialism.

Meanwhile, research tells us that kids with materialistic values are less likely to engage in actions that protect the environment.[22] And, to circle back to happiness, it turns out that both teenagers and adults who engage in environmentally friendly behaviors are happier than those who don't.[23] Less obvious are the harms to social and emotional wellbeing associated with the materialistic beliefs, values, and attributes embedded in all corporate marketing. It turns out that, even aside from environmental worries, materialism isn't good for us. I've already cited research showing that the things we buy don't make us happy in any kind of sustaining way. What's also true is that both children and adults who believe that things *do* make us happy are actually less happy than those who believe otherwise.[24] If you think about it, this makes perfect sense since the former are pinning their hopes for happiness on what's proven to be a false promise. Here's what does make us more sustainably happy than the things we buy: relationships[25] and intrinsically motivated experiences.[26]

Additionally, the harms of embracing materialistic values extend beyond their negative effect on happiness. Materialistic values are associated with compromised wellbeing and decreased satisfaction with life. In children, they are linked to depression, anxiety, lower self-esteem, psychosomatic illnesses,[27] underachievement in school,[28] and conflictual relationships with their parents.[29]

As spending on advertising increases, children become more materialistic. Compared with the 1970s, teenagers in the twenty-first century are significantly more likely to say that it's important to accrue lots of money and to have expensive possessions like a fancy car or a vacation home.[30] And materialism also seems to affect children's relationships with friends and classmates. In a 2015 study, a majority of children between the ages of eight and fifteen agreed that having "cool things" and "looking good" provide stepping-stones toward popularity.[31]

Advertising and commercial culture even influence young children's relationships. Preschoolers who watch a television program embedded with two commercials for the same toy and are then shown two pictures, one of a boy holding that same toy and one of a boy without the toy, prefer to play with the boy holding that toy—even if he's described as "not nice." Children who saw the program without the commercials were significantly more likely to prefer the boy who's described as "nice." In addition, kids who saw the commercials are more likely to prefer to play with that toy rather than play with their friends.[32]

Another study shows that even without seeing a particular product, brands have social significance for young children. Preschoolers who are familiar with brands use ownership of those brands as a measure of whether they think another child is fun as opposed to boring or popular instead of unpopular.[33] Even in early childhood, for those who internalize the messages of commercial culture, it seems that, as a character in the film *High Fidelity* observes, "What really matters is *what* you like, not what you *are* like."[34]

Research links children's materialistic values with the time they spend watching television or using social media.³⁵ But screens are not the only purveyors of commercial culture that celebrate materialism. Remember the song "My Favorite Things" from *The Sound of Music*? As a cure for being hurt or feeling sad, the lyrics suggest turning to simple pleasures like snowflakes, sleigh bells, kittens, wild geese, and delicious food. Now, however, there's a new version. For over eight weeks in 2019, the number one song on *Billboard*'s charts was "7 Rings," popstar Ariana Grande's twenty-first-century take on Rodgers and Hammerstein's 1959 classic. Here's what cheers up the former kids' TV star from the Nickelodeon series *Victorious*: diamonds, ATM machines, Tiffany's, champagne, credit cards, and writing checks.

According to the fashion magazine *Elle*, "7 Rings" is about Ariana Grande's resilience.³⁶ The lyrics of the song are rooted in a shopping trip she and her friends took to Tiffany's after a particularly trying time in her life, following a broken engagement and the death by drug overdose of a former boyfriend.³⁷ I originally thought the song was a parody or social critique, but it's actually a sincere paeon to retail therapy. ³⁸

> Whoever said money can't solve your problems
> Must not have had enough money to solve 'em
> They say, "which one?" I say, "Nah, I want 'em all."
> Happiness is the same price as red bottoms*

Speaking of retail therapy, the term seems to have first appeared in print in a 1986 *Chicago Tribune* article.³⁹ It may have initially been meant as a kind of satirical comment on the escalating me-first materialism of the 1980s. As evidenced by Ariana Grande's use of the term, however, it's a real thing. It's interesting that *retail*

---

* In case you're wondering, *red bottoms* refers to a distinctive hallmark of the popular Christian Louboutin line of shoes, which cost upward of $500 and are recognizable by their red soles.

therapy first appeared in the *Oxford English Dictionary* in 2001, the same year that, in the wake of the September 11 terrorist attacks, President George W. Bush famously told us to cope by going shopping.[40] Finally, in keeping with Urie Bronfenbrenner's concept that children are influenced not just by their family and community but also by seemingly distant societal forces, which I discuss in chapter 3, even children who aren't particularly materialistic themselves are harmed when they live in a society dominated by commercialism and materialistic values. When psychologist Tim Kasser, who has researched the psychological causes and effects of materialism for over twenty years, looked at the prevailing values in twenty wealthy nations, he found that the more a country's citizens prioritize possessions, wealth, power, and status, the more children are targeted with advertising on television and the lower that country ranks on the UNICEF scale of child wellbeing. On the other hand, countries that prioritize social justice, cooperation, and equality score higher on the UNICEF scale. Children in those countries see significantly less television advertising.[41]

I've been fortunate enough to work with Tim over the years, and I've learned a lot from him. So, I was interested in what he had to say about this particular study. He explained, "To me, this finding speaks volumes about what is really happening in society's attitudes toward children: If society cares about profit and money, then businesses and governments are likely to treat children as objects to be manipulated for profit, but if society cares about helping others, it recognizes that children need extra protection from those who would otherwise manipulate them, and it therefore limits advertising."[42]

As I wrote in *Consuming Kids*, all this talk about materialistic values makes me anxious. Especially since I've been immersed in thinking about the impact of marketing on children, I'm doing my best to be conscious of how my life reflects my priorities and values. I have to admit, however, that I'm embarrassingly vulnerable to certain products and the commercial messages that sell them.

I've developed a strange affinity for single-use kitchen gadgets, like mango slicers and avocado scooper-outers and even a banana slicer that cuts multiple slices simultaneously.

The commercials that most influence me change as I grow and develop. Two decades ago, I was haunted most by ones that preyed on my fears about my daughter's future, like the one featuring two guys talking about college—one who clearly has the money to pay his kid's tuition and one who doesn't because he didn't use an investment broker who asked the right questions. The last thing we see is the guy who screwed up by not choosing that investment firm: he is a picture of guilt, remorse, and anxiety.

Whether or not we have investment advisers, and whether those advisers are or are not asking the "right" questions, many of us identify with those feelings, especially given the skyrocketing costs of college. Many years later, I'm annoyed at myself for being drawn to ads for arthritis medications despite my conviction that it's wrong to advertise drugs on television. In other words, I make no claims of being unseduced by the material world. I do know, however, that a culture that immerses children in consumerism is doing a lousy job of teaching them to value what matters most: meaningful human relationships, love and kindness, awe and wonder, creativity, a connection with nature, and a deep appreciation for that which can't be packaged, bought, or sold.

5

How Rewarding Are Rewards?

If I offered you a thousand dollars to take off your shoes, you'd very likely accept—and then I could triumphantly announce that "rewards work." But as with punishments, they can never help someone develop a commitment *to a task or action, a reason to keep doing it when there's no longer a payoff.*

—Alfie Kohn, *Unconditional Parenting*

I am in Melbourne, Australia, snuggling on a couch with five-year-old Noah, who is patiently instructing me in how to use his Lego City racing app. Since we are only a few feet away from his impressive collection of actual Legos, I can't escape the irony that instead of constructing physical vehicles, towers, or what have you, we are bent over his tablet. "My mother-in-law bought it a few years ago when we were flying to see family in the U.S.," his father explains. "She thought it would make the flight easier."

Lego City apps are described on iTunes as "creative mission-based games."[1] Based on my experience with other Lego apps, I am not surprised to be underwhelmed by the opportunities for creativity afforded by this one. The digital cars are prebuilt and the (also prebuilt) accessories attach only one way. Nor am I surprised that we rack up points each time we "race" our cars. I am, however, taken aback when Noah announces excitedly, "Now we can go shopping!"

It turns out that the goal of this Lego app is not to build virtual cars, to race them, or even to accrue more points than your opponent. It's to use our points to buy virtual stuff, including a fire engine, a venus flytrap, and some de facto ads in the forms of a bus and a street sign advertising Lego's Ninjago empire, which includes the *Lego Ninjago Movie*, the *Lego Ninjago Movie Video Game*, *Masters of Spinjitzu* online games, the Cartoon Network's TV series *Lego Ninjago: Masters of Spinjitzu*, branded clothing, and more. We "buy" our prizes and start the process all over again.

Unlike many apps for children that require us to spend real money, we are only pretend shopping. But still, the app's message is clear: the fun of playing a game and the satisfaction of doing well, or even the pleasure in besting your opponent, aren't satisfying enough on their own. The real purpose of playing the game—its true importance—is to accrue points in order to buy things.

Parents looking for guidance in choosing among the zillions of apps, videos, films, and TV shows for young children can turn to organizations like Common Sense Media (CSM) and Children and Media Australia (CMA) for reviews and ratings of media and entertainment products for children. Common Sense Media, for example, lets parents know whether an app or program contains violence, sexualization, or references to drugs and alcohol as well as the amount of what they call "consumerism" involved (branded characters, in-app purchases, and other advertising). They also identify and rate the presence and effectiveness of the app's educational potential—the skills and information children could learn from them. A nonpaying membership to Common Sense Media allows for three free reviews a month. Parents must purchase a "plus" membership for unlimited access to reviews.

CMA's reviews are more detailed, particularly about the types of advertising and in-app purchases embedded in apps.[2] They are also free. In addition, CMA flags app and game features that might encourage gambling—like loot boxes. Loot boxes are virtual versions of the blind bags or boxes that I describe in chapter 4 that are used to keep kids searching for the one rare toy they need for their set of collectibles. In apps and games, loot boxes can appear as treasure chests, or other kinds of containers, filled with unknown contents and purchased with real money. The catch is that the loot box you buy may or may not contain what you want or need to succeed at the game. If a loot box does contain what you hope for, it's exciting and gratifying enough to incline you to buy the next one you encounter. And if it doesn't, there's always the possibility that the next loot box you buy *will* get you what you want.

The reviews I found on these sites and others are helpful. They don't, however, capture what I find most troubling about the Lego City racing app. They don't answer the crucial question: What does the program, game, or app teach children about what's most important in the world? The reviews don't talk about values—the fundamental guiding principles, goals, or motivations, that shape our life choices, our behavior, and our priorities and how we distinguish right from wrong.[3]

It's commonly accepted that children learn values from their families and from active affiliation with religious, cultural, and educational institutions, all of which are overtly committed to transmitting values from generation to generation. But children don't stop absorbing and adopting values the moment they leave their house, school, or place of worship.

Children also absorb values from other people they regularly spend time with—peers, neighbors, and other adults in their communities. Kids who spend hours each day with screens also absorb values from the owners and employees of the tech and media companies that manufacture the content kids consume on their phones, tablets, TVs, and other devices.

Because children usually experience tech and media creations in the absence of the people responsible for making them, it's easy for parents and caregivers to avoid conscious recognition of the values these creations communicate. However, every image, word, or idea encountered on a screen, or through a device, is selected by someone, or by an algorithm created by someone, who, perhaps in consort with lots of other people or at the command of their bosses, discarded a whole lot of other words, images, and ideas along the way.

In other words, what we do and do not encounter virtually, or in any medium, is a conscious or unconscious reflection of somebody's values—including their decision to compromise values to stay employed, to gain power, to make money, or to have a viable creative outlet. So, while kids don't spend time with these producers, writers, and app developers, they do spend a great deal of time exposed to, and perhaps absorbing, values held consciously or unconsciously by these people, which are reflected and transmitted through the content they produce and that children engage with.

It might make financial sense for companies whose profits depend on getting kids to spend money to train them to embrace consumption. But it doesn't make sense for children. It turns out that the materialistic values absorbed by kids immersed in commercialism affect their lives way beyond influencing how much and what they consume.

Beginning in the 1980s, a psychologist named Shalom Schwartz began conducting research in what grew to be eighty-two countries, identifying what he called a universal system of basic human values.[4] This doesn't mean that everyone prioritizes the same values, rather that there is a set of universally recognized values organized under an umbrella of discrete, inter-related categories. In addition, Schwartz and his colleagues found that within that system, certain values tend to be compatible—like conformity and security, while others are in conflict, such as benevolence and power.

That values tend to cluster together is relevant to more recent research on materialism. For instance, people who are primarily motivated by the materialistic goal of accumulating wealth and possessions are also likely to be motivated by values like power, status, fame, achievement, and popularity.[5] In other words, their actions and behaviors may be driven more by the external rewards rather than any intrinsic satisfaction derived from whatever they are doing.[6]

On the other hand, people who are less materialistic tend to prioritize values like egalitarianism, creativity, and concern about the wellbeing of others.[7] Values like these tend to motivate actions and behavior that may not bring material rewards but are satisfying in and of themselves.[8]

A question that comes up frequently in discussions about materialistic values is whether they are influenced by socioeconomic factors. Are people who struggle economically more materialistic than people who are well-off financially? It turns out studies do find an association between the insecurity of living in poverty and prioritizing money and possessions.[9] But one study that is particularly relevant to conversations about commercial culture found that *feeling* impoverished is linked to having materialistic values.[10] I think the word *feeling* is important in this context, and I wish there were more research on this.

It seems to me that (as long as you aren't struggling to meet basic needs) if everyone around you is more or less in the same financial boat, then it's certainly possible to have little money and not feel badly off. One problem with living in a society characterized by vast differences in wealth and with wide exposure to commercialized media and social networks is that, regardless of our financial situation, we are engulfed by images that glamorize, prioritize, and simultaneously normalize excessive consumption. We are continually reminded of what we can and cannot afford to buy and, by extension, how much money we have and don't have.

It makes sense to me that pervasive and ongoing exposure to marketing messages that trigger longing, and link happiness and self-worth to the things we buy would encourage people who struggle financially to embrace the extrinsic values and motivations associated with acquiring the wealth to buy those things. Yet, as I've already mentioned, believing that the things we buy can make us happy can actually diminish our happiness.

One reason people who primarily prioritize accumulating wealth and possessions may be perpetually dissatisfied is that in addition to mistakenly believing happiness lies in their next purchase, their sense of wellbeing depends on achieving a continual stream of external rewards, such as new purchases, compliments, applause, prizes, and grades. Further, in spending their time pursuing these rewards, they may have fewer experiences that truly promote their wellbeing.

For better or worse, at least in the United States and other industrialized countries, adults have long used material rewards to motivate children to engage in potentially distasteful but beneficial behaviors. Schools routinely use grades and other rewards to motivate students to study harder, learn new skills, and absorb information. Parents may base allowances on chores kids do around the house or give them money or presents as rewards for "good" grades or behavior. Rewards may motivate behaviors in the short term, but they can end up fostering materialistic values that persist into adulthood. Kids whose parents rely on material rewards are more likely to become adults who link their sense of self to what they own.[11]

In my experience, occasionally offering a material reward can help some kids overcome specific hurdles and navigate challenging terrain. When my daughter was in third grade, for instance, my husband and I resorted to offering an external reward to help motivate her to complete a school assignment that was difficult for her on multiple levels. She was required to memorize and recite some multiplication tables within a short amount of time. My daughter was, and always will be, someone who approaches tasks slowly and carefully. She found coping with the time constraint both frustrating and difficult. Because the task was counter to her nature, she kept avoiding practicing it. Meanwhile, she had a longtime wish for one particularly cuddly stuffed animal—a gorilla. So, we purchased the gorilla with the understanding that she could have it when she successfully completed her math challenge. Which she did.

It's problematic, however, when adults continually rely primarily on the promise of material rewards or punishments to manage or to change children's behavior in the long term. A significant body of research shows that material rewards aren't particularly effective for permanently changing or shaping behavior over time and that a steady diet of them may ultimately undermine children's healthy development.[12] Consistently paying children or buying them treats for helping out, for sharing, or for being kind, for instance, deprives them of opportunities to experience the satisfactions inherent in assuming the responsibilities, as well as the benefits, of being an integral member of a family, organization, or society.

Similarly, training children in school to prioritize getting "good" grades deprives kids of opportunities to experience the intrinsic gratification of learning—to satisfy curiosity, to gain competence, and to broaden and deepen their understanding of the world. When children learn to behave in certain ways solely to obtain a reward or to avoid a punishment, they miss out on opportunities to develop a sense of autonomy and to regulate their own behavior.

Another problem with relying on material rewards to motivate children to change their behavior is that we can't control every single thing kids learn from any particular experience. For instance, kids who repeatedly receive presents or money in exchange for behaving a certain way or completing a task may learn to do what their parents or teachers expect of them. It's also true, however, that they may simultaneously learn that things are only worth doing if they result in some form of payment, reward, or elevated status.

Children need lots of experience in environments that support and nurture their experience of intrinsic motivation. Until recently, kids were likely to fill whatever leisure time was available to them with initiating and carrying out their own internally motivated activities. That's when they could play, which is by definition the quintessential intrinsically motivated activity. Children don't play to get external rewards or treats. They play because they want to—for the sheer joy of it.

In our current tech-saturated, commercialized culture, we can no longer assume that children use their leisure time for self-directed, hands-on active or creative play. By 2009, before the advent of smartphones and tablets, watching television was the major leisure time activity for children around the world.[13] Today, of course, children spend increasing amounts of time with digital games on tablets and phones, where apps like Noah's Lego City racing game teach them that experiences, such as playing, learning, or doing something, are satisfying primarily if they result in some kind of external reward—even one that (a) exists only virtually and (b) can't actually be used in any way.

In contrast, when Noah plays with his actual Legos, he doesn't expect to go shopping when he's done or to get some kind of external reward, except perhaps his parents' expressed enjoyment in what he creates. His play with blocks is intrinsically motivated, perhaps because he enjoys building and/or because he has a vision of something he wants to create. He is intrinsically motivated to play, and his rewards are primarily internal: enjoyment in how he spends his time and the personal satisfaction of gaining skills or completing whatever building challenge he sets for himself.

Noah benefits from his intrinsically motivated building in all sorts of ways. He gets to experience himself as competent and independent. He learns that he can initiate a project and bring it to completion by himself. If he's building a structure of his own imagination, rather than working with a kit, his play nurtures his creativity. By linking "shopping" to doing well at the game, Lego's app makers diminish both the value of doing something because you enjoy it and the value of personal satisfaction in an achievement.

And speaking of external rewards, here's a joke my cousin told me: A psychologist, needing solitude and silence to write her masterpiece, rents a house on a street that, she discovers after moving in, is teeming with children who love to play loudly and exuberantly in her front yard. She asks them to play quietly. She asks them to play somewhere else. She puts up NO PLAYING signs. She resorts to yelling at them to stop. Nothing works.

One day, she tells the kids that she will pay them each a dollar to play as long as they like right next to her house. The kids are thrilled. And, true to her word, for the next several days they continue to play in front of her house and collect their dollars from her. One morning, she tells them, sadly, that she can no longer afford to pay them but that they are welcome to play in front of her house as much as they like. The kids walk away, silently, and never play in front of her house again.

On the one hand, this is merely a joke. On the other hand, it's a joke backed by research. Studies on motivation show that paying or otherwise materially rewarding kids (and grown-ups) for doing something they enjoy can diminish their intrinsic motivation to continue doing it once they are no longer paid for it. For instance, college students paid to solve puzzles were less likely to continue to work on them once they were no longer being paid. Students who were never paid continued to work on the puzzles because they were fun and challenging, even when they were told they could do other things.[14]

In another study, among preschool children who showed an interest in drawing, some were offered a reward for making a picture and some were just told to draw. Two weeks later, those who drew with an expectation of receiving a reward were less likely to engage in drawing than those who never received a reward in the first place.[15]

In a society permeated by advertising designed to sell consumers on the essential nature of consuming "stuff," belief in the power of external rewards and product-based panaceas can easily become a cultural norm. This phenomenon became abundantly clear during what I now think of as the "Great Pokémon GO Craze" from a few years ago. In chapter 1, I describe Pokémon's business model as both "brilliant and diabolical" because it is built on creating perpetually unfulfilled longing in children. It's worth taking a long look at Pokémon because it's an example of a product with a lot of intrinsic appeal to children but whose marketing exploits their vulnerabilities.

In the early summer of 2016, pedestrians around the world were treated to crowds of strangers, often bunched together, staring at their smartphones and occasionally crying out incomprehensibly "Raticate!" or "Clefable!" or "I've got a Vileplume!" They were playing Pokémon GO, the first wildly successful augmented reality game created by Niantic, a Google spinoff owned by its parent company Alphabet, and by Nintendo. The companies worked together to insert animated (and oddly named) Pokémon creatures into everyday life viewed through a smartphone.

Pokémon GO rapidly became an international phenomenon. It is available in more than 120 countries.[16] In just its first few months, it had five hundred million downloads and generated $470 million.[17] And it's no flash in the pan. Three years later, Pokémon GO had been downloaded over 1 billion times.[18] And from January to September, 2021, it was the fourth-highest earning mobile game and grossed a respectable $691.43 million.[19]

Upon its release, Pokémon GO was so incredibly popular that even the most vocal critics of marketing to children were hesitant to criticize it too harshly. Rather than call out Niantic for inserting virtual advertising for Pokémon in cemeteries, houses of worship, international monuments, and public parks,[20] some critics merely urged the company not to cash in by luring kids to McDonald's and other commercial enterprises.[21] Others focused on concerns about privacy.[22] And some metaphorically threw up their hands and praised the app for getting kids outside.[23]

Within weeks of its launch, myriad news stories featured quotes from physicians and other health experts lauding Pokémon GO for promoting exercise for couch potatoes.[24] Some even praised Niantic for tricking people into exercising.[25] CEOs of fitness tracker companies marveled at the astonishing rise of steps taken by people using the apps.[26] A headline in the *Wall Street Journal* proclaimed, "Want to Exercise More? Try Screen Time."[27] It wasn't until researchers tracked the effect of Pokémon GO on exercise over time that a different picture emerged. A study in the *British Medical Journal* found that while Pokémon GO players initially increased their number of daily steps, within six weeks they walked only as much as they had before downloading the app.[28] Other researchers found that the success of the app's contribution to maintaining a physically active lifestyle depended on how much users already valued exercise.[29] If players were walking primarily for the thrill of capturing Bulbasaurs and Butterfrees, then once their interest in the game waned, so did the positive health effects of Pokémon GO.

Since the goal of Pokémon GO was to generate profits and to promote brand loyalty, not to generate exercise and to promote public health, research debunking the app's potential as a healthy lifestyle tool was likely no big deal to its creators. At its inception, the game and its unproven but supposedly miraculous impact on exercise made headlines around the world. As often happens with headline-grabbing phenomena, the press lost interest and the news that the app was not a solution for curing couch potatoism received little, if any, mainstream coverage. So, I shouldn't have been surprised to learn that the company plans to release a brand-new Pokémon app with an alleged benefit to humanity featured front and center in its marketing strategy. The new app is called Pokémon Sleep. And, you guessed it, the app is supposed to reward "good sleeping habits."[30] Since sleep is most definitely a biologically driven and, by definition, an intrinsically rewarding function of all humans and animals, I was eager to learn more how this app is going to work.

At a press conference, Tsunekazu Ishihara, CEO of the Pokémon Company, announced that "the concept of this game is for players to look forward to waking up every morning."[31] Then he added, "Everyone spends a large part of their lives sleeping, *and turning that into entertainment* [my italics] is our next challenge." Wow.

As of this writing, the Pokémon Company has not yet spelled out exactly how sleep is going to be rewarded, except to say that what happens in the game is going to be tied to when and how long a person sleeps.[32] Given the huge financial success of Pokémon GO, Pokémon Sleep is likely to be simultaneously highly lucrative in the first few weeks after its release and highly unlikely to help most people sleep.

Based on research debunking Pokémon GO's influence on exercise, what we know about sleep, and the problems discussed earlier in this chapter of relying on external rewards to effect sustained behavior changes, it's hard to see Pokémon Sleep as anything other than a cynical effort to monetize a legitimate public health worry. It's certainly true that many children and adolescents get significantly less sleep than they need.[33] It's also true, however, that time with tech and even sleeping in the same room with a television or digital device are documented factors in sleep deprivation.[34]

I am particularly puzzled by Mr. Ishihara's claim that Pokémon Sleep will somehow get people to "look forward to waking up in the morning." In my experience, people who do not look forward to waking up are likely suffering from depression, grief, or extreme stress, all of which may require professional help and none of which are likely to be helped by Pikachu or any Pokémon creature. Besides, sleep experts recommend getting in the habit of clearing our minds before going to sleep and to avoid focusing on either future or past events.[35] So, dwelling too much either positively or negatively on waking up in the morning could prevent people from falling asleep.

Meanwhile, except for the naturally occurring phenomenon of dreaming, sleep, by its very definition, shouldn't and can't be entertaining. According to the *Oxford English Dictionary*, sleep is defined as "a natural and unconscious state or condition universally assumed by people and animals during which the activity of the nervous system is almost or entirely suspended, and recuperation of its powers takes place."[36]

The Pokémon Company's proposed transformation of sleep into money-making entertainment did not spring full blown out of nowhere. It's one more step in the tech and marketing industry's efforts to gamify and brand all aspects of our lives. As a *New York Times* headline put so aptly, "Pokémon Sleep Wants to Make Snoozing a Game Too."[37]

The app was supposed to be released in 2020, but as of this writing that still hasn't happened. Believe me, I'm not losing sleep over the delay. The company did, however, release an app called Pokémon Smile. According to the Pokémon Smile website, "Getting kids to brush their teeth can be tough, but this clever app turns toothbrushing into an exciting adventure!"[38] Kids brush their teeth while staring into a smartphone at themselves in a virtual Pokémon hat. As they brush, they can save Pokémon creatures from "harmful bacteria," and then they get to catch all the Pokémon they've saved. The narrator assures the viewer that the game will "motivate players to brush and play every day."[39] It was only on my fourth viewing of the video that I noticed the tiny disclaimer that appeared at the bottom of the screen for all of six seconds: "This app is not intended to prevent or treat cavities. Nor does it guarantee that players will gain a liking for toothbrushing or make it a habit."[40] Yet don't the words "motivate," "brush," and "every day" presented in the same sentence suggest otherwise?

In case you're still not convinced about the company's cynical take on health, Nabisco recently released packages of sugar-filled Pokémon Oreos. Just as the lure of purchasing Pokémon cards rests on the possibility of collecting rare ones, the same is true of packages of these creature-bearing Oreos—which just might possibly contain the elusive Mew. When I heard that these edible and extremely crumbly Mews were selling for $1,000 on eBay, I decided to take a look. In fact, the prices ranged from $6.00 to a whopping $25,000 (no takers the last time I checked). The first one I found for $1,000 was listed as "Collectible Mew Oreo Cookie Limited Edition Cracked." (No bids for this one either, last time I checked.)[41] But one intact Mew Oreo had a high bid of $13,300—but hey, half the proceeds went to St. Jude Children's Research Hospital. It's a worthy cause, but I can't help but wonder why it hasn't seemed to occur to the bidders that they could save $6,650 and just donate the money directly to the hospital.[42] As a commentator on NPR asked, "How long do Oreos last?"[43] But I digress—back to toothbrushing.

I couldn't find any research on the effectiveness of Pokémon Smile, but I did find a study suggesting that over the course of three months, preschoolers using a reward-based tooth-brushing app did seem to do a more effective job of brushing than kids who did not use the app. However, as the researchers themselves point out, the app's true effectiveness can only be measured in a long-term study. In addition, this study shares a fallacy with many of the other studies I've seen on screen-based interventions. The comparison is almost always to a control group that gets no intervention of any kind. So what we really end up learning is that something, possibly anything, is better than nothing.[44]

I don't have anything against trying to make potentially tedious tasks fun for young children—although I doubt that anything can make tooth brushing an "exciting adventure" in any kind of sustained way. When my daughter was first learning to brush her teeth, we made funny noises together as she brushed: "eee . . . ahhh . . . ohh" and that kind of thing. Or sometimes we brushed our teeth together. My objection is to corporate attempts to monetize these tasks of daily living by using their brands as motivation.

Perhaps you remember one of the more egregious of these: CTA Digital's iPotty with Activity Seat for iPad. Back in 2013, extrinsically motivating kids with time on an iPad was Big Tech's solution for toilet training. The iPotty won Fairplay's Worst Toy of the Year award that year, and I'm happy to say it appears to no longer be manufactured.[45]

Whenever I spend time with young children, particularly babies and toddlers, I am reminded that the capacity for intrinsic motivation is deeply rooted in the human condition. In *The Case for Make Believe*, I wrote about the time I was lucky enough to be visiting a friend at the moment his seven-month-old daughter made an astounding discovery—her knees. Squealing with glee, she extended her arms to her father, expressing in no uncertain terms her desire to stand up. As each tiny fist gripped tightly to one of his fingers, she pushed up from her toes and straightened to a standing position. After a few wobbly upright moments, she began to squat, bending her legs slowly. Then, like an inebriated ballerina rising from a plié, she teetered up once more. Beaming with pride, she repeated the sequence again—and again and again and again.

Eventually she noticed a favorite toy kitten on the floor. Holding on with only one hand, wobbling even more ferociously, she began to reach for the kitten, only to find that (a) it was too far away to grab and (b) it was at ground level. With great deliberation, she extended her free hand toward the cat. Tottering precariously, completely focused on her mission, she began the glorious process of bending—and was saved from an undignified tumble by her father's protective arm. She allowed herself a brief rest on the floor and, with joyful determination, began the process anew.

In nurturing environments where there are opportunities to explore the world on their own terms, young children are intrinsically motivated to learn, to gain competence, to strive for autonomy, and to satisfy their curiosity. It's true that most parents I know have occasionally relied on a reward to help their children get through an onerous experience or a task for which their kids have absolutely no motivation. But when we immerse kids in experience after experience built on a system of external rewards, we not only prime them to be motivated only, or mostly, by material gains, we also deprive them of opportunities to experience the deep fulfillment that comes from doing something—anything—because they care deeply about it.

6

The Nagging Problem of Pester Power

The circle of retail life. Child demands product, parent learns about product through child, household begins using product, child ideally grows up to encourage his or her own household to use said product—at least until their own kids start making requests.
  —David Sprinkle, research director
    of the market research firm Packaged Facts

I am chatting with parents at a local preschool when the mother of a three-year-old says, "My parents were babysitting and showed my daughter these videos where kids open up boxes of toys. Now she keeps bugging me to buy them and she can't understand why I won't." A father nods and sighs, "I try not to take my son to toy stores, but now *every* store seems to have toys for sale. I went to Home Depot for light bulbs and there were toys even there! We had to leave before I could buy what I needed because he kept nagging me for stuff and had a full-out tantrum when I said no."

Coping with nagging is one of the more unpleasant, frustrating, and, when it culminates in public tantrums, embarrassing challenges of parenting a young child. It's a common source of stress for families.[1] Unfortunately, getting kids to pester their parents is also a proven money-maker for corporations. And there's no question that, around the world, advertising directly to kids influences the things children pester parents to buy.[2] At one family gathering in New York, for instance, my cousin had this to say about his five-year-old daughter, whose life had been completely free of commercial television—until a few days before: "She'd never asked for anything until then," he said. "And then she wanted every single toy that was advertised!" A year earlier at another family gathering, I heard from a nephew whose son is a few years older. "First it was Minecraft, and now it's Fortnite," he says. "My eleven-year-old begs constantly to play games online. And once he starts, I can't get him to stop. I know he's playing too much, but I just get worn down." In other words, targeting kids with advertising encourages them to nag.[3] And nagging works.[4]

Given that the journey from infancy to adulthood is a slow transition from helplessness to relative autonomy, it's not surprising that many children at least attempt to pester their parents to get what they want. Even very young children experience desires equally as intense and sometimes as mercurial as any adult longings. Unlike adults, however, kids don't have the freedom, the cognitive capabilities, the physical strength, or the cash to get what they want when an adult refuses a request.

A poll conducted in 2002 suggests that the persistence with which children nag seems to increase as they get older. A survey of 750 kids between the ages of twelve and seventeen found that, on average, they might ask nine times before their parents give in and let them have what they want. Nagging seems to peak in early adolescence. Of the twelve- and thirteen-year-olds surveyed, 11 percent reported nagging parents more than fifty times for one specific product or another—and all of these were products they had seen advertised.[5]

Nor is it surprising that many parents have a hard time saying no to nagging—even when we know it would be better not to give in. We succumb for all sorts of reasons, not the least of which is the fundamental truth that we want our children to be happy.

Or maybe we're already saying no to a bunch of other requests. Or we're ambivalent about whatever it is that our kids want. Or we're stressed about work and it's easier to avoid a struggle. Or we feel guilty. And so on. In a study of mothers in the United States, researchers found that most of them viewed nagging as a conflict and many described it as a "battle."[6]

Marketers didn't invent nagging, but they certainly exacerbate the problem. It's bizarre and cruel that companies not only count on pestering to bolster sales of everything from candy to cars but encourage kids to do it, while absolving themselves of responsibility for nagging-related family stress. Instead, companies place the blame on "very permissive, cooperative, and too-busy-to-matter" parents who "underwrite" nagging and then "blame it on the marketplace."[7] British marketing expert Elliott Haworth is typical: "Advertisers aren't forcing parents to buy a Big Mac with a gallon of coke for their nagging children. I'm sure it's annoying when a child screams for sugar, but the (in)ability to say 'no' is a reflection of ineffective parenting rather than advertising being a malicious force."[8]

I don't absolve parents of responsibility for their children's behavior in a commercially driven world, but the possibility that parents aren't coping well doesn't justify marketers' efforts to get kids to nag. Most of the parents I talk to do their best in what can feel like an unending and overwhelming struggle. In the face of well-funded, brilliantly strategized, and ubiquitous commercial assaults on their children, it's unfair to expect parents to be both gatekeepers and children's sole protectors.

There are also marketers who see their work as serving to neutralize parents as gatekeepers. When Haworth interviewed Emma Worrollo, head of the Pineapple Lounge, an international market research firm specializing in kids, she dismissed the idea of parents shielding children from advertisers, advising fellow marketers to get around it.*

---

*Emma Worrollo is mistakenly identified as "Holly Worrollo" in the article.

"I hate the word 'gatekeepers' . . . that's not how you should be thinking of a person you want to engage with your brand. Loads of brands have used that strategy—talking to parents only. But now we're in a place where it is possible to have a dual strategy: it's possible to create an experience that will talk to different generations. Families now are more like teams and it's less about something for the parents, something for the kids—there's a more symbiotic happiness that we can create."[9]

Worrollo's characterization of families as teams exemplifies the marketing industry's penchant for exploiting currently accepted theories of child development, social psychology, and family dynamics. Encouraging children to think of their family as a team is a common and potentially useful strategy for helping them to experience their family as a cohesive unit thriving on attributes such as cooperation and loyalty and mutual goal setting. For Worrollo, however, the concept of family teams is justification for targeting kids, as well as parents, with marketing, even as she ignores the developmental vulnerabilities that make young children so vulnerable to marketing manipulation.

The same article complains that advertising to parents instead of children diminishes children's "choice as consumers."[10] This reminds me of an interviewer who once asked me, "Don't kids have a right to be marketed to?" I believe strongly that children have many important, and even crucial, rights. According to the United Nations Convention on the Rights of the Child, kids have the right to free expression, to play, to live free of race-based discrimination, and more. Nowhere does it say that children have a right to be targets for marketing.[11] The convention does assign children the right to live free of exploitation, which—given that marketers routinely exploit children's vulnerabilities—could certainly include living free of child-targeted advertising.

It's become so common to count on boosting sales through kids' nagging that marketers even coined nicknames for the phenomenon. It's called "pester power" or, less frequently, the "nag factor."

The term pester power in connection with advertising to children surfaced in a 1979 article in the *Washington Post* about consumer activists advocating that the Federal Trade Commission ban television advertising to young children.[12] Their efforts failed,* and by the 1990s, marketers were unabashed in and unapologetic for their efforts to increase profits by getting kids to nag. A one-day conference held in London was called Pester Power: How to Reach Kids in 1994.[13]

Four years later, a market research firm called Western Media International published a study on nagging titled "The Nag Factor."[14] It remains one of the most telling examples of the marketing industry's indifference to the wellbeing of children and families. Its purpose was not to help parents cope with nagging or to reduce children's nagging. It was designed to help corporations provide tools to help kids become more successful naggers.†

"The Nag Factor" study identified two kinds of nagging, which they called "persistence nagging" and "importance nagging."[15] As Cheryl Idell, director of strategic planning and research at Western Media International explained, "It's not just getting kids to whine, it's giving them a specific reason to ask for the product."[16]

Western Media's researchers asked 150 mothers of young children to track their kids' nagging over a two-week period. In total, the mothers recorded a total of ten thousand nags, which comes to about sixty-seven nags per child.[17] The study found that nagging accounted for as much as 46 percent of sales for companies that

---

*While the U.S. effort was unsuccessful, similar bans were passed in Sweden, Norway, and the province of Quebec in the 1980s and 1990s. Norwegian Ministry of Culture, Broadcasting Act no. 127, December 4, 1992, https://www.regjerin gen.no/en/dokumenter/broadcasting-act-/id420612/; Brandon Mitchener, "Sweden Encourages Rest of Europe to Restrict Children's Advertising," *Wall Street Journal*, May 22, 2001; Bill Jeffrey, "The Supreme Court of Canada's Appraisal of the 1980 Ban on Advertising to Children in Quebec: Implication for 'Misleading' Advertising Elsewhere," *Loyola of Los Angeles Law Review* 39, no. 1 (2006): 237–76.

† I describe this study in detail in chapter 2 of *Consuming Kids*.

target children,[18] which translates into a whole lot of cash. According to a 2018 report by the media conglomerate Viacom,* which at the time owned Nickelodeon, MTV, Paramount Pictures, and more,[19] young children and preteens "wield $1.2 trillion in annual purchasing power in direct and indirect spending."[20] According to a survey included in the report, 77 percent of kids ask for products they see in commercials—and 74 percent of parents accede to their requests.[21]

Not surprisingly, the 1998 nag factor study found that the more stressed parents are, the more likely their kids are to be successful naggers. The parents least likely to give in to nagging? The ones who are most affluent and those unlikely to have babies or toddlers in the house.[22]

Aside from embodying the marketing industry's disregard for the welfare of children and families, the nag factor study lays bare a significant problem with market research. In academia, research is regulated by the Human Subjects Review. Academic researchers must inform potential subjects about the research and any risks involved in participating. In the United States, at least, market research is subject to no such regulation. Imagine a researcher approaching you and saying, "Will you participate in a study to find out how to help your kids become more effective naggers?" Would you participate?

When I was interviewed for the 2003 film *The Corporation*, I was asked about the most egregious practices of marketing to kids. Encouraging kids to nag was high on my list. The film's director tracked down Lucy Hughes, vice-president of Initiative at Media, one of the companies sponsoring the study, and interviewed her as well. She justified the nag factor study this way: "If we understand what motivates a parent to buy a product . . . if we could develop a creative commercial—you know a 30-second commercial that

---

*Viacom merged with CBS in 2019 and is now called ViacomCBS.

encourages the child to whine ... that the child understands and is able to reiterate to the parents, then we're successful."[23]

When Ms. Hughes is asked about the ethics of encouraging kids to nag, she says, "Is it ethical? I don't know. But our role at Initiative is to move products. And if we know you move products with a certain creative execution placed in a certain type of media vehicle, then we've done our job."[24] By encouraging children to nag, one aspect of "doing our job" is to encourage consumerism by fomenting family stress.

In a way, the nag factor study was a gift to those of us mobilizing to counter the escalation of marketing to children. *The Corporation* was the first of a spate of films, books, articles, and other media highlighting the nag factor as emblematic about why children should not be targets for marketing.[25] Shining a light on the fact that corporate marketers "intentionally try to make parents lives absolutely miserable"[26] by getting kids to nag was shocking enough to help us easily poke holes in the prevailing marketing industry stance that "steps to stop children from seeing advertising are completely unnecessary" because "advertising to children is an entirely harmless activity."[27]

In the United States, the advertising industry responded to the bad publicity by expunging terms like "pester power" and the "nag factor" from their public statements. When I attend marketing conferences now, I find that marketers here rarely speak overtly about getting kids to nag. Instead, they bend over backward to use benign but insidious phrases like "collaborative decision making" and child-to-parent "communication" about products.

Way back in 2007, however, the European Union banned advertising that overtly urged children to pester their parents for products.[28] That may have stopped companies from creating ads explicitly asking kids to nag their parents, but it ignores the problem that any ad designed to create desire and longing in children implicitly encourages them to nag.

In the United States, in 2009, the Children's Advertising Review Unit (CARU), which is the advertising industry's self-regulatory arm, took a similar tack when it addressed nagging in its guidelines: "Advertising should not urge children to ask parents or others to buy products. It should not suggest that a parent or adult who purchases a product or service for a child is better, or intelligent or more generous than one who does not."[29] The problem is that (a) CARU has no regulatory power over advertisers and (b) again, regardless of whether an ad tells kids to nag overtly, immersing them in ads expertly designed to create longing is bound to result in requests that can easily morph into nagging. Four years later, two commercials for Nintendo's Wii made clear the ineffectiveness of relying on guidelines to curb advertising to children. In each commercial, children practice "importance nagging" by presenting hapless parents with a list of reasons why they should buy the Wii.[30]

Today, in developing nations where targeting children with advertising is a more recent occurrence and where resistance to marketing is less organized and less vocal, marketers still don't seem to feel the need to disguise their intent. In the past few years, I've seen studies on kids' marketing-induced nagging in countries as diverse as Pakistan, Greece, Iran, and South Africa—and on every continent but Antarctica.

A study of Iranian children published in 2015 in the *British Food Journal* draws the unsurprising conclusion that packaging affects children's desire for products and that kids' nagging during visits to a grocery store influences their parents' purchases.[31] What is most notable about this particular study is the authors' conclusion that their findings are important because they will help marketers devise "effective ways of influencing children's purchasing decisions with the aim of getting children to more powerfully influence the purchasing decisions of their parents through the use of different forms of pester power such as crying and begging." The authors conclude by telling companies to "present their products in ways that are especially appealing to children."[32]

In India, researchers exhort marketers to target children because "today's ultragenerational young child is very smart in forming collaboration with siblings [and], grandparents to pester their parents."[33] Others describe how kids in India "use different pestering strategies such as negotiation, bargaining, requesting, threatening, crying, and often shameful repetition to achieve what they want."[34]

Marketers targeting kids in developing countries tend to be more upfront about monetizing nagging. A market analyst in South Africa claims that nagging is so powerful that children are responsible for 40 percent of parents' choice of automobile color. Another finds that children aged eight to fourteen from eleven countries are able to influence their family's choice of automobile 60 percent of the time.[35] And speaking of cars, an executive from the Publicis Groupe, the largest advertising conglomerate in the world, describes an ad for a ŠKODA vehicle that was marketed in India this way: "The biggest challenge was to do a pure product film in the tonality of brand ŠKODA. Weaving a family story around the product features does the job nicely for us. The added benefit obviously is the pester power we appeal to that is playing increasingly an important role in big purchases in any home."[36]

In Pakistan, the author of an article titled "The Payoff in Marketing to Kids" (spoiler alert: it's not family harmony), explains that advertising to children is "all about creating pester power, because advertisers know what a powerful force it can be." She adds that research is terribly important because "marketers must know exactly what makes kids tick. They should have in-depth knowledge about children's behavior, emotional and social needs at different ages, etc."[37]

Around the world, mounting evidence shows that marketing-related nagging is harmful to children and families. It's a factor in family conflict, overspending, and debt. In Australia, market research shows that children and parents average one conflict every seven minutes when shopping at a grocery store.[38] A 2018 survey of one thousand parents in Scotland found that more than a third of parents actually go into debt to satisfy their children's nagging.[39] A mom in England writes about the stress of trying to limit her sons' time playing Fortnite, which has been called the most successful and addictive video game ever.[40] She describes parents hiding game consoles in the refrigerator or in their underwear drawers and going through all sorts of contortions because they can't seem to withstand their kids' nagging.[41]

During my stint on a call-in radio show, a baby boomer tells me that "it's all the parents' fault. They are too indulgent these days. They need to learn to say no." I frequently hear comments like this when I give talks about children and the marketplace, mostly from those in the audience whose children are grown. But after years of exploring advertising and advertising practices as they affect children, I've come to the following conclusion: telling parents to "just say no" to every marketing-related request they feel is unsafe, unaffordable, unreasonable, or contrary to family values is about as simplistic and ineffective as telling a drug addict to "just say no" to drugs.

And it's getting harder. It was difficult enough for parents when children's exposure to marketing consisted mainly of thirty-second television commercials for toys linked to popular television programs or movies. Today, families cope with popular apps and games for children purposely designed by tech companies to be habit forming.[42] A father who limits his eleven-year-old son's screen time told me that "it's like he's a different kid once he starts playing online. He begs me to get on, and then begs not to get off when his hour is up. I hold him to the limits we set, but I hate that always he pushes me for more time." And not only do apps and online games keep children online for huge amounts of time, many of the most popular apps and games for kids assault them with all kinds of inescapable advertising.

Researchers at the University of Michigan looked at 135 of the most commonly downloaded free and paid apps in the Google Play Store for children five and under and found that 95 percent of them contained some kind of advertising.[43] All the free apps had ads, but paying for an app did not make it ad-free. A startling 88 percent of paid apps also contained advertising. Nor are apps labeled "educational" any less filled with sneaky advertising than other apps.

Almost two-thirds of the "free" games include what are called "full-app teasers," ads to upgrade to the paid or "full" version of the app. In the app Balloon Pop, to give just one example, the fancier or more interesting balloons can be popped only in the paid version. They appear on the screen in the free version so that kids are reminded that the paid version is so much better every time they try to pop one. Almost half the apps are de facto ads for other products in that they feature commercial icons, like the animals from *Paw Patrol* or Lego characters. Other common marketing techniques include alluring come-ons to spend money to get to a higher level, to purchase extras to decorate and personalize avatars, to continue playing, or to have a more fun and varied experience.

Pediatrician Jenny Radesky, lead researcher on the University of Michigan study, was deeply disturbed by the findings. "The first word that comes to mind is furious," she told the *New York Times*. "I'm a researcher. I want to stay objective. . . . My frustrated response is about all the surprising, potentially deceptive stuff we found."[44]

What also makes life harder for parents and children today is that, at least in part because of lack of public funding for public institutions, the lines between commerce and everything else have gotten so blurred that children are targets for marketing in places where we least expect it. I went to the Museum of Science in Boston with a cousin and her seven-year-old twins recently. She wisely told them ahead of time that they could go into the gift shop and look but that she was not going to buy them anything there. What they weren't prepared for, however, is that the museum had another gift shop, seamlessly integrated into the train exhibit, and filled with especially appealing toys. The twins were enraptured and—of course—longed for all sorts of things. She held her ground, and the kids accepted her decision, but it added unnecessary stress to the afternoon. In a similar vein, a friend told me about visiting the Museum of Fine Arts in Boston after Thanksgiving, just in time for a big holiday market in the museum's courtyard. She bumped into a mom coping with a ten-year-old who was asking to buy some of the jewelry on display. "I thought a trip to the museum was going to be a respite from commercialism" the mom sighed. "Clearly I was wrong."

Marketing to children has become so pervasive, ubiquitous, invasive, and seductive that it competes with parents to capture children's hearts, minds, and souls. Whether kids encounter it on television, embedded in apps, or when they are out and about in the world, marketing contributes to stress in families,[45] and marketers know that quite well.[46] That knowledge, however, doesn't stop them either from targeting kids or from encouraging others to do it.

At a marketing conference I attended, a keynote focused on the "emotional states" and "passion points" of millennial families today. According to the Pew Charitable Trusts, millennials were born between 1980 and 1997.[47] In 2004, when *Consuming Kids* was published, they were the kids inundated with marketing. Now their children are targets.

The keynote speaker was George Carey, CEO of a market research firm called the Family Room. The Family Room's mission is to "elevate the role of brands in the life of millennial families by speaking to the family's heart, not their heads [sic]."[48] As he speaks to us about the emotional state of millennial families, I begin to worry. The Family Room helps companies operationalize the advice I first heard at a different marketing conference, this one on artificial intelligence and emotion: marketers should get with the "emotion economy." Companies should no longer tout the quality of their products or use advertising to explain what their products do. Instead, they should capture and exploit our emotions. They need to monetize our feelings.

Carey exhorted those of us in the audience to remember that the feelings of millennial families "have incredible importance for those of us in the commercial space trying to create emotionally relevant solutions for kids and families."[49] Whether companies are selling cars or kids toys or snacks or media entertainment, they must keep track of "solutions" that are "emotionally relevant" to families. "Emotionally relevant solutions" is market-speak for advertising that works. In his pitch to us, an audience mostly of prospective clients, the speaker set out to explain how the emotional states of millennial families "are embedded in your brands, or your character, or your platform, or your breakfast cereal."

He neither asked nor answered the most crucial question: Is it really in the best interest of children or families to have their emotional states embedded in brands like Nike, or in characters like Spiderman, or in platforms like YouTube Kids, or in cereal like Honey Nut Cheerios? Probably not. Millennial families are experiencing the highest poverty rates of any young families in twenty-five years—more than one in every five millennial families lives in poverty.[50] They are the generation most affected by the Great Recession of 2008, so it can't be good for their emotional wellbeing to be embedded in what they buy.

We learned that what millennial parents want most is time—time to be with their children. This is consistent with social science research on millennial families showing that they value a work-life balance, but that such a balance is more difficult than ever to achieve. About four million millennial families are headed by a single parent.[51] Of millennial families with two parents, 78 percent—more than any other generation—have both parents working at least full-time.[52] And even before COVID, millennials felt more obligated than other generations to work from home even after their official workday is over.[53]

Given that we know children benefit from face-to-face time with their parents or guardians, the scarcity of millennial family time is a problem that deserves our attention. As a society, we should have a thoughtful discussion about possible solutions, including, but not limited to, raising the minimum wage, allowing for flexible work schedules, and ensuring that employees aren't obligated to work after hours. Instead, the solution presented to the marketers in the audience is that they need to create programming that the whole family can enjoy, which increases the likelihood that millennial parents and children spend their precious family leisure time with all eyes on [add your brand here].

We also learned that millennial parents are so child-centric that they forgo outside friendships. Carey tells us that almost 70 percent of mothers in the United States and almost all the mothers in China say that their child is their best friend. He prefaces his next point—perhaps the most important point he is making—with an aside. "I know," he says, "that there are some developmental psychologists in the room who would tell us that this is not a good thing. But this is where we're headed, and these numbers keep going up every year."

What was merely an aside for Carey points to one of the most harmful prevailing marketing philosophies. In their endless quest for advertising opportunities, marketers exploit the ways society fails children and families. Marketing directly to children exacerbates an ongoing, normal tension in family life that arises as children move from the total dependence of infancy to the relative independence of adulthood.

Finally, Carey built to what may be, from his perspective, the most crucial information he had to share: that parents and children are now best friends has implications for how families make choices because "you're not going to boss your best friend around when it comes on decision-making." Only one in four moms in the United States describes herself as the family decision maker, down from 85 percent seven years ago. More than half of moms with children ages six to seventeen, 56 percent, say that family decision-making is now led by kids, with parental guidance.[54] He continued, "People have a sense of families as hierarchies, but I'm here to tell you that it's over. It's not a hierarchy—it's a web. And how you market to, and entertain, a family web is different from how you market to a hierarchy. It's a different kettle of fish." In other words, marketing to children is as important, or more important, than marketing to adults.

How parents and children navigate this terrain depends on factors such as the child's temperament, the cast of characters in the immediate and extended family, and the parents' own temperaments, culture, religion, and values. Carey didn't offer advice for parents, but experts that do frequently offer the admonishment "pick your battles." This used to be good advice. But now that children are subjected to myriad, finely honed marketing campaigns for everything from junk food to junk apps, it's almost impossible for parents to know which battles to pick.

Marketing experts and others point to problems related to single-parent families, two-career families, lack of adequate day care, and all sorts of other reasons for stress between parents and children. I agree that many of these issues are major concerns. I also believe that neither the marketing industry nor anyone else has the right to take advantage of children whose parents are unable to provide ideal care.

Except for the fact that children and families are being harmed, there's something darkly comic about living in a commercialized culture that thrives on business models dependent on encouraging obnoxious behavior in children. No sane parent would welcome people in their home whose every interaction with children is designed to instill in kids such intense desires that they nag incessantly to get them fulfilled. Yet that's exactly the goal of all advertising to children. In one study, more than one-third of the moms interviewed felt that limiting commercial media exposure reduced nagging.[55]

A slew of parenting blogs are filled with suggestions about how to cope with and forestall all children's nagging. These generally focus on what parents can do in their own families, but they rarely address the necessity of limiting the child-targeted marketing that is often at the root of parent-child conflicts over what and how much to buy.

7

Divisive Devices

I'm hopeful that the next generation of children, after watching us be fools for our devices, may decide it's not worth it.
—Jenny Radesky, MD, University of Michigan Medical School, author of numerous studies on children and technology

I am walking through a park near my house after a storm and happen on a toddler lugging an enormous branch. His mom is sitting on a bench a few feet away, intent on her phone. "Mom!" he calls as he inches closer to her. His mom, head down and staring at her screen, doesn't look up. "Uh-huh," she says. He and the branch laboriously edge closer. "Mom!" he calls, with more urgency. "Uh-huh?" she answers, still staring at her phone. I am transfixed, wondering what will happen next. Clearly frustrated, he gets even closer and yells even more loudly, "MOM!" She still doesn't look up and so . . . he slugs her! Clearly, it's the one thing he can think of guaranteed to get her attention. And it does, although it's probably not the kind of attention he was hoping for.

Public discourse about the effect of tech-delivered commercialism on children's wellbeing used to focus primarily on either the amount of time they spent on various devices or on whatever content they were consuming. But the advent of smartphones changed all that. Today, it's not just children's overuse of technology that makes headlines. So does overuse by parents and caregivers. There's mounting evidence that adults' vulnerability to the lure of persuasive design places a whole new strain on their interactions and relationships with their kids, which are so central to children's wellbeing.*

I don't believe in corporal punishment for parents, or children, but I do admit that in the scene I describe above, my sympathies are with the toddler. And yet, I also recognize that—as much as we love them, and as much as they bring us joy—there are times when being the parent of a very young child can be boring, stressful, and sometimes both.

When my daughter was young, before the ubiquity of smartphones and tablets, I played with her, of course, but certainly not all the time. There were plenty of times when we were in proximity, but she was playing by herself. And that's when I sat in the same room, or on the edge of the sandbox at a nearby park, and read. I always took a book with me to the park. And so when parents are criticized for being on their phones too much, I sometimes wonder what I would do if I had young children today. I know what I *hope* I'd do, but I was never faced with that dilemma. And given that I read—a lot—when taking care of my daughter, it feels hypocritical for me to be too judgmental. Judgment aside, however, it's increasingly clear that the tech industry's success in capturing and holding their users' attention can interfere with crucial family relationships.

In fact, there's mounting evidence that the mom and her branch-toting toddler I witnessed in the park—she bent over her phone, and he increasingly frustrated by his unsuccessful attempts to get her to look at him—are not unique. Research suggests that

---

*There's a growing body of literature on the impact of digital device use on adult interactions and relationships as well, but that is not the focus of this book. See for instance, Sherry Turkle's *Reclaiming Conversation*.

when absorbed in their devices, parents are more apt to ignore children's bids for attention than when they pass the time in device-free ways,[1] like chatting with other adults while their kids play independently.[2] Also worrisome—when parents immersed in a device do respond, it's more likely to be with hostility.[3]

Of course, the parents researchers observed who were off their devices didn't necessarily accede to their kids' requests.[4] What's most important is that they at least acknowledged that their child *made* one. In my experience, and as my colleagues who do couples or family therapy tell me, ignoring someone you're close to escalates tension by inculcating frustration, if not rage, in the person whose existence is, at least momentarily, being denied. No one likes being ignored, and children are no exception. And current research does suggest a link between parents' device use and young children's problematic behavior, ranging from tantrums to withdrawal.[5] One possible conclusion is that it's a vicious cycle. Parents stressed by young children's behavior may escape into their devices, and when kids can't get their parents' attention, they express their frustration by acting out more.[6] Like most toddlers, the little boy who slugged his mom in the park didn't have the language to say, "I'm hurt and angry that you won't pay attention to me" or the self-control to keep from physicalizing his anger.

So is there any difference between me reading a book while my daughter played and parents today scrolling through their phones while their children play? Certainly, our intent is the same—to pass the time or to escape momentarily from the stresses of daily life. But the words printed on the pages of a book are static—they don't change depending on the vulnerabilities of whoever is reading. They don't move around, beep, or flash at us to get our attention, and we don't anticipate instantaneous, external rewards for staying focused. Losing oneself in a world evoked by reading is absorbing, but the medium delivering that world—the book itself—is not designed to foment addiction. "But what about reading on a device?" a skeptical friend asks when I test out this theory on her.

"Isn't that the same thing?" I do find that reading for pleasure on a device that doesn't feature ready access to alluring distractions like email, hyperlinks, texts, or social media does provide an experience similar to reading a book (especially since my e-reader's case mimics the heft of a one). I, however, find it difficult to read anything long or substantive on my phone, although other people do read long articles and even books on theirs. They tend, however, to read them either early in the morning or late at night,[7] which suggests that parents in the throes of taking care of young children are more likely to be texting, emailing, playing a game, or scrolling through social media and therefore are more likely to be captured by variable rewards, personalized content, and advertising, as well as other persuasive techniques.

I was unable to find a study comparing the ease with which adults disengage from a book compared with from a smartphone. I did, however, find evidence that toddlers are more prone to tantrums when *they* transition off being on a screen than when they transition from engaging with a book.[8] And, just like adults glued to their phones, preschool children immersed in screens seem less likely to respond to their parents' requests for attention than when they look at books or play with analog toys.[9] This isn't surprising since, like apps for adults, apps for children employ features designed especially to capture and hold their attention. For kids, these features include bright colors, autoplay, and beloved media characters.[10]

Toddlers, preschoolers, and parents aren't the only family members whose immersion in screens can shut out the people around them. The same is true for older kids and adolescents. Teens' and preteens' obsession with, or addiction to, video games, social media, and other forms of tech, is an escalating family problem.[11] Adding to the tension is that parents feel guilty about the time they and their children spend with tech. And they often feel powerless to do anything about it.[12]

In my more cynical moments, I find myself thinking of the tech and media industries as purposely wreaking havoc not only on families but on all our close ties to other people. After all, the more we neglect to prioritize our intimate human relationships, the more dependent we become on our devices for comfort and companionship. And the more time we spend on our devices, the more money we generate for tech companies and their corporate clients. In my less cynical moments, I think the best word to describe the creators of our devices—at least in relationship to much of the content they create—is amoral, not immoral. Harming people is not the goal of their business plans. Instead, the prize they have their eyes on is beating out competition for our attention, and thereby generating the greatest possible profit. As for potential harms, well, people can choose when, how, and how much to use their devices.

Those of us concerned about the effect of too much tech time on children often point to media stories about tech executives who send their kids to screen-free schools and/or set limits in their own homes. More than anyone, the people creating these technologies must be familiar with the potential harms to children. It was the late Steve Jobs, who when asked whether his kids love the iPad replied, "They haven't used it." He then explained, "We limit how much technology our kids use at home."[13] But I see another, more subtle, message. If, the argument goes, the people making the technologies protect their children from using—or overusing—them, then shouldn't, or couldn't, other parents set limits as well?

The flaw in this argument has to do with two messages implied in Steve Jobs's admission that he kept his kids away from the iPad. The first is, "if I can protect my kids, then so can everyone else." The other is, "the burden on protecting kids should be placed solely on parents, not on my company."

In fact, laying the blame and burden solely on parents is absurdly simplistic. As Tristan Harris, co-founder of the Center for Humane Technology, says about self-control and the lure of our digital devices, "You could say that it's my responsibility . . . but that's not acknowledging that there's a thousand people on the other side of the screen whose job is to break down whatever responsibility I can maintain."[14] And don't forget the billions of dollars riding on their success. Facebook alone generated $86 billion in 2020, more than ten times its revenue in 2013, most of which comes from advertising.[15]

Jaron Lanier, often called the "father of virtual reality," put it even more bluntly: "It's important to remember that these devices are designed to be addictive, in the formal sense. That's the acknowledged truth from executives at these companies. Rather than thinking it's the kids' fault or the parents' fault, we have to recognize that these are cruel systems that prey on universal human frailties."[16]

More than fifteen years ago, I was at a marketing conference and attended a session on cell phones, where the speaker made the following point: The adult cell phone market was becoming saturated and, like all markets, it needed to expand. The teen market was also getting saturated, and so the next target was going to be preteens, and then even younger kids. The speaker suggested marketing phones as safety devices to parents and as "fun" or "cool" to kids. It continues to be a remarkably effective strategy. Fifty-three percent of eleven-year-olds in the United States own a smartphone. So do 32 percent of ten-year-olds and 19 percent of eight-year-olds.[17] In 2005, I was especially worried about the impact of these portable devices on children's play, creativity, and opportunities to engage in the world around them. More than a decade and a half later, I now understand it also as an assault on the important bonding experience of conversations between parents and children.

When my daughter was in elementary school, her friends' moms and I used to talk about the strategy of waiting until a car ride to have potentially difficult talks with our kids. The intimacy of being close together in an small space, but not face-to-face, seemed to lend itself to conversation. Around 2005, as video content became more readily available on cell phones, media companies began to see car rides as a potentially lucrative opportunity. As I wrote in *The Case for Make Believe*, when Verizon snagged a contract with Sesame Workshop to make *Sesame Street* content available on its phone service, J. Paul Marcum, then head of Sesame Workshop's interactive division, denied that the company advocated marketing cell phones to young children. Then he added, "But you can't ignore the convenience factor when people are in motion. A parent can pass back a telephone to the kids in the back of the car. And it's a device that families are going to carry with them everywhere."[18] Around the same time, a spokesperson for Verizon said, "Parents who have some time with a kid are finding [video downloads on cell phones] to be a great diversion."[19] Ken Heyer, a market researcher for ABI Technologies, put it this way in the *New York Times*: "It's really convenient because there's only so much 'I Spy' that you can play."[20]

And here's where we are today. Whether with reluctance or open arms, we have invited into our homes powerful, seductive entities designed to generate profits by monopolizing our attention. And they don't give a damn about our family relations or our children's wellbeing. Unlike a cigar, which is sometimes just a cigar, our smartphones are not just machines. They are conduits for tech companies to surveil and mold our behavior—kind of like alluring, fun, charming, endlessly engrossing, and benign-seeming corporate spies who are expert at insinuating themselves into our lives and manipulating us to their own ends. We take them to dinner, on vacations, to the park with our kids, and to bed. We ask their advice, rely on them for information, and count on them to calm our kids down or to keep them entertained. While appearing to serve us, the real mission for smartphones is to serve us and our children up to advertisers to generate corporate profits. In other words, corporations have figured out how to infiltrate and monetize family life, and we pay them to do it.

Beginning in the twentieth century, as communication technologies evolved, corporations found increasingly powerful ways to insinuate themselves into our homes and our families. In the 1930s, advertising-supported newspapers and magazines migrated from being hawked on the streets to being delivered to doorsteps.[21] Around the same time, radio-delivered, advertising-supported news and entertainment, which reached even more people because it had no literacy requirements, began to monopolize family time and heralded the tech-enabled spread of consumer culture in the United States. By 1941, two-thirds of radio programs had advertising.[22] Families sitting around and listening had less need to rely on themselves for home entertainment and amusements. And the time they spent listening to the radio was time they weren't spending talking, playing games, or making music together. Starting in the 1950s, as Robert Putnam eloquently describes in *Bowling Alone*, television had profound influence on isolating us from our communities.[23]

The new technologies, however, with their capacities for surveillance, their persuasive design, their addictive properties, their brilliant marketing, and their capacity to give us what we think we want, disrupt families more than any previous technologies have. Although smartphones are the most visible and ubiquitous of the digital intruders into our relationships, they aren't the only ones. Parents talk less to children when they read enhanced e-books and play with electronic toys together.[24] The flourishes e-books can bring to text—images that move and talk when tapped or swiped—and the noises generated from electronic toys fill silences that might otherwise be filled with parents and kids talking together. While this reduced conversation is often cited as a concern about decreasing language acquisition, it's also another example of how devices come between parents and children. We often cuddle with young children when we read books to them. The combination of physical intimacy and shared experience is calming for both parents and children and can be a springboard for wide-ranging conversations that strengthen our connection. It turns out, however, that parents and children not only talk less together when reading on a tablet, they are also less apt to cuddle.[25] And this seems to hold true even when the device they use has no digital enhancements.

In her book *Reclaiming Conversation: The Power of Talk in a Digital Age*, psychologist Sherry Turkle highlights the importance of conversation for deepening relationships. "In family conversations," she says, "children learn what can matter most is not the information shared, but the relationship sustained."[26] I find myself reflecting on Turkle's quote a lot as I think about the Internet of Things—so-called smart objects that connect to the internet. These can be everything from toasters to vibrators to toys for children.[27] They are marketed as desirable because they "learn" our preferences and habits by tracking what we do with them, analyzing the data they collect and altering their behavior accordingly. They can also share what they've learned and collect even more data by "talking" to our smartphones, tablets, computers, and other nearby IoT objects.

The most obvious problem with any smart device is invasion of privacy. They collect untold amounts of information about us and our kids. On the one hand, the data these devices collect can be used to personalize their services, making them even more attractive to us. On the other hand, the data can also be used to personalize advertising, making it even more effective. But when it comes to the potential to disrupt or to weaken family bonds, I worry more about digital personal assistants, like Amazon's Alexa, particularly when they target children.

Voice-activated digital assistants and their "search" capacity have become increasingly popular, ubiquitous, and invasive. According to a blog from a software acquisitions company called Ignite Technologies:

> digital assistants won't just know you at home or in the office or when you use your smartphone, they will know you everywhere. You'll find your digital assistant in your watch, in the dashboard of your car, in your text messages,

in your fridge, in your TV, in your computer at work—everywhere. This has always been the endgame: an omnipresent digital assistant that knows you as well as you know yourself—and then some.[28]

While Google and Apple embed their digital assistants in their devices, Amazon leads the market in smart speakers, which connect to its digital assistant, Alexa.[29] Amazon markets Alexa to us as a combination servant, DJ, and savant. It's home to tens of thousands of apps, which Amazon calls "skills." Alexa can control your lights, lock your doors, set alarms, operate other smart devices, stream music, answer questions, and more.[30]

In contrast, Amazon markets Alexa to potential skills developers as a cash cow. Businesses are urged to invest in Alexa skills to "close the distance between your brand and your customers, and make money by selling premium experiences and physical goods."[31]

Amazon launched Alexa and its smart speaker, the Echo, in November 2014.[32] A kids' edition came out four years later—essentially the same speaker in a colorful protective case bundled with a subscription service with skills especially for children and parental controls.[33]

Apple and Google got a head start on Amazon in cultivating future brand-loyal customers with their extremely successful efforts to target kids in schools. In fact, the Echo Dot Kids Edition isn't the first, or only, Amazon device aimed at kids. Since 2014, Amazon has been trying to develop a direct connection to children. The company has introduced several versions of kids-edition tablets[34] and a kids-edition Kindle e-reader and has created a buying program specifically for tweens and teens.[35] As one reviewer said, the child-friendly Echo Dot is "a clear effort by Amazon to get kids to use its voice assistant instead of the Google assistant or Apple's Siri."[36] It's Big Tech's version of a battle for lifetime brand loyalty.

A huge problem with the Echo Dot Kids Edition, according to a 2019 complaint to the Federal Trade Commission (FTC) filed by nineteen advocacy groups, is that it violates children's privacy. The complaint states that the kids' version of the Echo Dot "has the capacity to collect vast amounts of sensitive personal information from children under the age of 13."[37] It "records children's voices any time it hears the 'wake' word* and uses artificial intelligence to respond to a child's request. Amazon stores these recordings in the cloud unless or until a parent deletes them."[38] Before the complaint was filed, Amazon kept the information it obtained from a child, even after the recording was deleted.[39]

In addition, Amazon tracks "how a child uses Echo and can use that information to recommend other content that the child might enjoy. Amazon may collect other types of personal information when children ask the Echo to remember something or when a kid skill solicits an open-ended communication from a child."[40] At the time, only about 15 percent of skills for kids had links to privacy policies.[41]

The advocates argue that the Echo Dot for kids violates the Children's Online Privacy Protection Act (COPPA), one of the few legal constraints in the United States protecting children's information from being exploited by companies that collect that data online. It's unfortunate that the FTC took no action on the complaint. Amazon made some changes, but it did not adequately address the advocates' concerns. Now, when parents delete recordings of a child's interaction with Alexa, both the recording and the information itself are deleted, but the burden is on parents to remember to delete their children's data instead of the device routinely deleting it when it is no longer needed.[42]

Given that Amazon has the potential to save children's data indefinitely and that kids may use the Echo Dot (or future versions of it or other smart Amazon products) into adulthood, Amazon has the potential to capture just about a lifetime of information about individuals. Proponents of digital assistants might respond

---

*The "wake" word is the word designated to turn on the device.

that the device's longevity and acquired knowledge make it increasingly useful because it can tailor its content to fit different interests. On the other hand, such knowledge in the hands of a giant, largely unregulated corporation only strengthens the possibilities for, and effectiveness of, exploitation in the form of behavioral advertising—in this case, ads based on previous information collected from young users that are designed to appeal especially to their particular desires.

There's no doubt that protecting children's privacy is crucial to preventing their exploitation, but that's not the only serious concern we should have about the Echo Dot Kids Edition. Amazon advertises that the Echo Dot will unlock "a world of kid-friendly content"[43] through Amazon Kid+ (free the first year and $3.99 per month after that), including "thousands of audible books, interactive games, and educational apps."[44] In addition, the Echo Dot has a drop-in feature, like an intercom, through which parents can communicate with their child from another room in the house—as long as a family owns another Echo device. It comes with parental controls that can be adjusted to set some limits on the content children can access, such as explicit songs. Parents can also set time limits and use the Amazon Parent Dashboard to review what their children actually do with the Echo Dot.[45] Amazon's marketing positions the Echo Dot as helping "kids learn and grow,"[46] which is described as, "Kids can ask Alexa questions, set alarms, and get help with homework."[47] So what's the problem?

In fact, there are many. From the perspective of Amazon's bottom line, I suspect that "learn and grow"[48] also means learning to love Amazon and growing to depend on it. Neal Shenoy, CEO and co-founder of BEGIN, an early learning tech company, was upfront about his goal for personal assistants when he said, "Tech has struggled with kids' pronunciations and fluency, but imagine a world where kids—even preschoolers who haven't learned to read—can ask their devices about facts and feelings. . . . It can be a *co-parent* [my italics] and help children learn everything."[49]

Do we really want corporations like Amazon training young children to depend on them to answer questions, tell stories, play games, sing lullabies, or help with homework? These are tasks that traditionally are an important component of caring for children and have long been performed by parents, other adult friends and relatives, librarians, or teachers—people whose primary interest in a child is not financial gain but that child's wellbeing.

Among the many problems with ceding child-care responsibilities to the predictive algorithms that select Alexa's responses to children is that they are based on what users like or what they have learned about users through their online and (increasingly as smart devices proliferate) offline behaviors and activities. At worst, these algorithms lead us down rabbit holes where the only information they get is what we want to hear. But even something as seemingly benign as basing children's book recommendations on what they've previously liked isn't necessarily the best thing for them. It precludes opportunities for kids to expand their horizons and their comprehension of how the world works. Recommendations from commercially driven algorithms aren't made from a well of love, or a deep understanding of a child's needs, or hopes for that child's future that a parent, a teacher, or a local librarian might have.

And speaking of librarians, when I was a child, my sister Nancy and I spent a great deal of time at our local library. My sister, who was and still is a voracious reader, developed a close relationship with the children's librarian, Miss Whitehead,* who took an interest in what my sister was reading. Like many girls, she went through a long horse phase when all she wanted to read was books with horses. Miss Whitehead, who thought it might be good for my sister to expand her reading experiences, enticed her to other books this way: "Nancy," she would say, "wouldn't you like to be the first person in this library to read {insert name of a horse book here}?" Nancy, of course, would. "Well," Miss Whitehead

---

*This was before the universal *Ms.* was introduced.

continued, "I'll save it for you. But first you have to try {insert name of non-horse book here}." Which my sister did. That kind of, well, bribery might not work for some kids, but Miss Whitehead—who had known her for a long time at that point—was right when she thought it would work for my sister.

In addition, sharing stories, songs, and games from our own childhoods can deepen family bonds. They provide kids with the security of experiencing themselves as being in relationships based on long histories and provide parents with an opportunity to pass on their heritage. As I was working on this chapter, my adult daughter texted a picture of a bridge over a stream in the woods. "This reminds me of something," she texted. I knew instantly what she was remembering. "Poohsticks!" I replied. It's a game in A.A. Milne's *The House at Pooh Corner* that involves dropping sticks off a bridge into a stream. Whichever stick makes it to the other side of the bridge first wins. We played Poohsticks when she was a child, and I played it with my family when I was growing up.[50]

Of course, lots of young children won't immediately experience the significance of family history, but it's likely they will as adults. And the continuity of relationships can have meaning for some kids even in the moment. During one of my virtual puppet chats with kids when so much was shut down because of COVID, six-year-old Theodora decided to read Audrey a story. She picked up a book about trolls and monsters and explained, "My father had this book when he was a little boy." Clearly, the book's connection to her dad made it special.

Amazon's advertising for the Echo Dot touts its affiliations with entertainment conglomerates like Disney, Nickelodeon, and Warner Brothers, as well as with commercialized icons like Barbie and SpongeBob SquarePants, whose images are licensed to sell toys, food, clothing, and accessories to children.[51] The more I looked into the Echo Dot for kids, the more I wondered how its commercial partnerships would affect its recommendations for kids. So, I bought one, signing up an imaginary four-year-old girl. While this may seem young, Amazon recommends the Echo Dot Kids Edition for kids beginning at the age of three[52] and allows parents to designate content for children as young as two.[53]

A few minutes into my first interaction with Alexa—before I made any brand-specific request—it asked if I wanted to hear an American girl story about someone named Julie Albright. Who, you might wonder, is Julie Albright? Turns out she's not just any American girl. She's one of Mattel's American Girl dolls, released in 2007 and updated in 2014.[54] The Julie doll, with a book about the character, sells for $110 on the American Girl website. Furniture and accessories are also available in prices ranging from $195 for "Julie's Groovy Bathroom" to $150 for her pinball machine. Various outfits go for $30.00 each. The only item for less than $10.00 is another book.[55] In addition, kids can watch Julie Albright movies, available on Amazon[56] and YouTube.[57]

Next, I decided to use the "Alexa, I'm bored" feature to see what the Echo Dot had to offer. "Would you like to play a game?" Alexa asked brightly. "Yes," I replied. "Star Wars Sound Journeys?" it asked. "No," I answered. It then proceeded to ask in succession if I want to try Disney Remix, the SpongeBob SquarePants Challenge, Barbie You Can Be Anything, and the Wizarding World Book Quiz.* After I rejected all those, Alexa went silent. I tried the next day and got the same five suggestions. Then I tried several times in succession. Sometimes I got the same ones, and sometimes the list of five included American Girl Adventures, Boss Baby (a 2017 computer animation film by DreamWorks), and *The Unlucky Adventures of Classroom Thirteen*, a book series published by Little, Brown and Company. We could argue about whether any of these suggestions are appropriate for a four-year-old, but there

---

*According to Wikipedia, Wizarding World is a fantasy media franchise and shared fictional universe centered on a series of films, based on the Harry Potter novel series by J.K. Rowling.

are even more concerning issues about Alexa recommending them to me in my faux four-year-old mode.

We don't know, because corporate algorithms are proprietary, how the algorithms selecting their content make their choices. Is it like other search engines in that what comes up is at least influenced by how much and whether a particular company pays for the privilege? Or whether the search engine owns that company?[58] Does content from companies that pay more get recommended more frequently or placed earlier in the queue? When it comes to the best interest of children, shouldn't Amazon be transparent about the answers to these questions?

Amazon promised that Alexa would be ad-free,[59] but what the company seemed to mean is that there would not be traditional advertising like commercials on television or radio. But my experience of having five commercial properties foisted on me by just telling Alexa that "I'm bored" is an awful lot like advertising within movies, television, and video games—either product placement (the insertion of a product into the content of a program) or host-selling (when the host of a program promotes a product within the content of a program). In fact, host-selling was made illegal on children's television programming in the United States in 1974 and is still not allowed on broadcast, cable, or satellite.[60] Yet, Alexa clearly functions as "host" of programming on the Echo Dot, since it's the personified conduit for the algorithms selecting any content streamed. It's hard to see why Alexa is allowed to direct children to branded content like the SpongeBob SquarePants Challenge when that practice isn't allowed on other media.

In fact, companies wanting to monetize Alexa and other voice assistants are encouraged to create content that will be useful to users while promoting their product. In 2018, a business blog titled "Amazon Alexa: How to Leverage the Benefits for Your Brand," explained that "experts believe that product placement and recommendations on Amazon Alexa and other AI platforms are a matter of time. Meanwhile, they already recommend that brands should have an associated Alexa skill."[61] Many already do. Kids can warble along with Frozen Sing! and learn to speak like Chewbacca, the loveable *Star Wars* Wookie, by asking Alexa to open C-3PO Translates![62] And so much more.

At the risk of sounding melodramatic, what especially worries me about the Echo Dot (4th Gen) Kids Edition is the form the hardware takes. Earlier versions of the Echo Dot for kids looked pretty much like the Echo Dot for adults—a miniature speaker—except the adult ones were black and the children's were brightly colored. The fourth edition, however, comes as either a round, chubby, adorable tiger face or an equally adorable panda. It's no accident that they are cute as hell. Humans are suckers for cuteness. It's hardwired in our brains, perhaps, to keep us perpetually in love with babies, despite the challenges of caring for them, thereby ensuring the propagation of our species.[63] Designers of robots' physical appearance are encouraged to build in cuteness to evoke engagement and a heartwarming response to these machines.[64] As Sherry Turkle puts it, "It's not that we have really invented machines that love us or care about us in any way, shape or form . . . but that we are ready to believe that they do."[65]

By transforming the Echo Dot from an obvious machine to what appears to be a loveable talking animal, Amazon has strengthened the likelihood that both children and adults will develop emotional attachments to these devices. Embodied in the little orange tiger face sitting unplugged on my kitchen table is the already rising tide of another tech tsunami: communicative robots, or "sociable bots," who use facsimiles of human conversation to generate bonding relationships with people.

Doing justice to the moral, ethical, and social complexities of the role of robots in our lives is beyond the scope of this book. But I do want to point out the dangerous possibility of corporations using these bots to exploit kids for profit—in the same vein as Alexa's unsolicited recommendations that I play games with brands owned by Disney, DreamWorks, Viacom (which owns

Nickelodeon), and Mattel (which owns Barbie). Take Hello Barbie, for instance, Mattel's first foray into manufacturing toys designed to hold personalized conversations with children. Several of the doll's possible responses include direct and indirect references to the brand's other toys.[66]

While commercialized digital technologies may interfere with children's family relationships, they strengthen what are called "parasocial" relationships. For the purposes of our discussion, parasocial relationships describe children's sense of intimacy or relationship with robots and screen-based characters. Research suggests that children can develop feelings for and strong attachments to robots. Kids also ascribe feelings and an inner life to these mechanical creatures, even though they know they aren't really alive.[67]

Meanwhile, the powerful and influential role beloved characters play in the lives of children is well documented. Putting well-known cartoon icons like Shrek and SpongeBob SquarePants on packaging can influence children's food choices.[68] As digital technologies become increasingly more sophisticated, researchers are looking at children's parasocial interactions. These interactions with screen-based puppets and animated characters generate what are called "socially contingent" utterances in direct response to what a child says, thereby mimicking actual human conversation.

One 2020 study, for instance, found that children interacting with a trusted and well-known screen character like Dora the Explorer learned a math skill faster than kids in a control group. And the more children liked and trusted Dora, the better they did with the task at hand.[69] The study concludes, "This interactive dimension will pave the way for children to embrace and trust media characters as effective intelligent social partners and teachers in the twenty-first century, as well as push the boundaries of what children will perceive to be alive."[70]

The researchers' conclusion is emblematic of a central problem with a great deal of the research on children and tech. The study evaluates the messages, or learning possibilities, of digital technologies as if they exist outside the problematic business plans that drive the products being evaluated. Nickelodeon's Dora the Explorer is a wonderful character, but she's also the front for a major commercial assault on kids.[71] Should kids (or parents) trust Dora if children's conversations with her were ever to be recorded by Paramount Global, Nickelodeon's parent company, and monetized? Is it ethical to encourage kids to "trust" Dora, when her image is used to sell kids myriad toys, food, clothing, accessories, and other forms of media?

As a ventriloquist who has spent decades engaging kids in parasocial relationships with my puppets, most recently through video chats, I've experienced the joy these relationships can bring. I have no doubt that children's parasocial relationships with engaging media characters either on a screen or through a robot-like object could facilitate all sorts of positive benefits. But here's what's troubling. As artificial intelligence improves and machine-driven characters become capable of increasingly complex conversations, kids' attachment to them will become stronger. And the potential for unregulated tech and media companies to exploit these attachments for financial gain will also increase—not, as I've said before, in the service of a child's wellbeing or best interest.

8

Bias for Sale

Prejudice, bias, stereotyping, and stigma are built not only into many games, but other forms of identity representations in social networks, virtual worlds, and more. These have real world effects on how we see ourselves and each other.

—D. Fox Harrell, professor of digital
media and artificial intelligence, MIT

In 2011, I received an unsettling email from a Fairplay supporter. "Why," she asked, "does an ad for Your Baby Can Read? pop up when I Google your name?" I had no idea, but it bothered me. Your Baby Can Read was a system of videos and flashcards claiming, as its name implied, to "teach babies to read." It was marketed on YouTube, Twitter, and Facebook and through television infomercials and ads on network and cable stations.[1]

Fairplay and our attorneys at Georgetown University's Institute for Public Representation had recently filed a complaint to the Federal Trade Commission to stop the company's false advertising claims. Among other press, the *Today* show had recently done a segment about our campaign, and we were hoping that, regardless of what the FTC did, we might prevent parents from believing the ads and spending significant amounts of cash on a product marketed with false claims.

It turned out the company had purchased the search term "Susan Linn" from Google in order to have their ad pop up every time someone Googled it.[2] Having my name enable advertising for a company whose mission I found abhorrent was deeply disturbing. I got in touch with someone I knew at Google and, after looking into it, he reported that there was nothing I could do about it. My experience served as a painfully personal awakening to fact that for-profit technologies as apparently dispassionate as a search engine can and do generate harm. In the scheme of things, of course, my experience with Your Baby Can Read was not much more than a personal irritant. It turns out, however, that allowing companies to purchase our names as keyword search terms can have much wider and more destructive consequences, for instance for people whose names may suggest they are Black.

Two years later, a study in the journal *ACM* Queue* showed that ads implying that a person has been arrested were more likely to appear in searches for names associated with being Black than for names associated with being white. For instance, a Google search for "Latanya Sweeney" turned up an ad for access to public records headlined "Latanya Sweeney, Arrested?" A search for "Kristin Linquist" turned up ads for public records, but they all had more neutral headlines, such as "We Found Kristin Linquist." This was true even though there were absolutely no arrest records associated with the name "Latanya Sweeney" (who happens to be the author of this study), yet there were arrest records associated with the name "Kristin Linquist."[3] This kind of embedded racism inflicts myriad potential harms to adults, especially for job applicants whose potential employers conduct online searches for information about prospects. But there are harms for kids as well, particularly the children of adults whose job prospects are derailed

---

* Association of Computing Machinery.

because of such searches, but also the children who for whatever reason Google their own name and find it—or names associated with their particular race—associated with being a criminal.

We shouldn't underestimate the power search engines have over children's knowledge of the world and how it works. Instead of turning to public institutions that are the traditional repositories of knowledge—libraries and schools—we and our children now rely on the algorithms of private, for-profit tech companies to answer our questions about the world. Today, that means our searches for information are dominated by just one company: Google.

In August 2021, a staggering 92,525 Google searches were conducted *in just one single second*.[4] About nine out of ten online searches are conducted on Google,[5] capturing close to 90 percent of search engine revenue.[6] Disturbingly, 66 percent of search engine users believe that the searches they conduct result in fair and unbiased sources of information. Among users eighteen to twenty-nine, the number of believers rises to 72 percent.[7]

Of course, the information provided by libraries and schools is by no means bias-free. Like all of us, the librarians, teachers, and administrators in charge of these institutions have conscious and unconscious biases, and these likely affect the information they and their institutions choose to provide. One important difference from Google, however, is transparency. Public institutions, and the people who work for them, can be held accountable for the decisions they make. In addition, the process by which those decisions are made is, or should be, transparent. The tech companies that create and own algorithms driving search engines and other online platforms have no legal obligation to share how or why the information they present is selected.[8]

Because search engine results appear on our devices with magic-like rapidity and because they are calculated by machines, the results seem divorced from human bias. It's important, however, to remember that, like librarians and teachers, the people who create algorithms also have biases that influence their creations.

What's more, instead of working for institutions whose primary mission is to serve the public interest, tech company employees fulfill their companies' primary goal by ensuring they make money. Just as Facebook isn't a public square, Google isn't a public library.

For instance, when we use Google Search, criteria for where a particular website appears in our results could include, among others, what Google "thinks" we want to see, based on our search history, other online activities, and any other data Google has collected about us.[9] The results also depend on techniques employed by search engine optimization (SEO) companies, a multibillion-dollar industry designed to help businesses secure top spots in internet searches. Paying for SEO does not come cheap, which makes search results likely to bias large companies rather than small businesses or cash-strapped nongovernmental organizations.[10] Meanwhile, since Google's algorithms are proprietary, we can't know exactly what influences which sources get priority in a search, so it's hard to assess the validity of what we encounter.

When we search for information on Google, the first results we see are usually ads, identified as such, from companies that have successfully bid to have their ad linked to whatever search term we enter. In 2020, Google generated about $147 billion from advertising revenue, most of it generated by auctioning off search terms to potential advertisers.[11] But, as I describe in the previous paragraph, the "organic search results," the ones that follow the ads and are supposed to provide the information we're looking for, are not prioritized objectively. And they may well be serving Google's economic interests. According to the *Wall Street Journal:* "Google has made algorithmic changes to its search results that favor big businesses over smaller ones, and in at least one case made changes on behalf of a major advertiser, eBay Inc., contrary to its public position that it never takes that type of action. The company also boosts some major websites, such as Amazon and Facebook, according to people familiar with the matter."[12]

Safiya Umoja Noble puts it more bluntly in her eye-opening

book *Algorithms of Oppression*, "Google biases search to its own economic interests for its profitability and to bolster its market dominance at any expense."[13] Noble's impetus to explore the ways Google's search engine algorithms perpetuate racism is rooted in her 2011 experience of using the search term "Black girls" and discovering that the first results were for pornography—not just in the ads associated with the search term but in the organic search results as well.[14]

It was not an isolated problem. Four years later, for instance, social media and the mainstream press reported that the top result in a Google search for "three Black teenagers" was criminal mug shots. Meanwhile, results from a search for "three white teenagers" turned up benign photos of white kids.[15] One of the most disturbing stories illustrating the potential dangers of search engines valuing profit over accuracy occurred the year before, in 2015, when Dylann Roof murdered nine African Americans at a Sunday church service in Charleston, South Carolina. He claimed in court and elsewhere that the path leading him to that horrific act began when he was a teenager and did a Google search for "Black on white crime." The first sites he encountered were not, for instance, police or FBI statistics on crime. Instead, what rose to the top of his search were sites filled with white supremacist propaganda and, as he told investigators, "That was it."[16]

In response to public outcry, Google fixed these harmful results and others, but their Wac-a-Mole approach utterly fails to address the larger issue. In 2020, a watchdog group reported that the tool Google offers to help companies decide what search terms they should link their ads to was still linking the phrase "Black girls" (as well as "Asian girls" and "Latina girls") to pornographic search terms.[17] The company's official stance is that racist search results are not a Google problem—they merely reflect society's biases.[18] Really? Given Google's enormous popularity and the fact it markets itself to schools as an educational platform, can't we assume that its results influence society's biases as well as reflect them?

Jessica Guynn, writing in *USA Today* about the "three Black teenagers" results, framed the problem this way:

> With cuts to public education and increased reliance on technology to deliver answers, search engines wield more power than ever before in deciding what information is seen and what information is important. People view Google as an unassailable source of credible and reliable information. Yet, what's often missing in search results that are not curated by a thoughtful hand, say a librarian or teacher: awareness of gender stereotypes and racial biases.[19]

Like so many companies responding to George Floyd's murder and the ensuing Black Lives Matter protests, Google put out a statement touting its commitment to "racial equity" and detailing its plan to achieve it. The plan included admirable goals like increasing minority hiring, committing funds to antibias education, and creating products that will be useful to the Black community. Missing from the statement, however, is any mention of how Google will ensure going forward that its search algorithms do not perpetuate or encourage racism.[20]

Meredith Broussard, an AI researcher at New York University, said it so well in the *New York Times*: "Computers are excellent at doing math, but math is not a social system. And algorithmic systems repeatedly fail at making social decisions. Algorithms can't effectively monitor or detect hate speech, they can't replace social workers in public assistance programs, they can't predict crime, they can't determine which job applicants are more suited than others, they can't do effective facial recognition, and they can't grade essays or replace teachers."[21] I completely agree with Broussard. On the other hand, as evidenced by the changes Google made to search results for "Black girls," algorithms can be tweaked to correct for racial bias, which suggests that such bias can be prevented in the first place.[22]

While writing the above, I remembered that cute little tiger-faced Echo Dot Kids Edition still sitting on my kitchen table. Since Amazon promotes Alexa's search feature as being able to answer children's questions and help with homework, I decided to try asking Alexa a question similar to Noble's Google search. Of course, it wasn't the same since Alexa seems to have no response to statements, only commands. "Alexa," I asked, "what are Black girls?" I admit it's not the most elegant phrasing, but it was the best I could come up with at the time. The only answer I got was that *Black Girl* was the name of a play by J.E. Franklin. So I tried again.

"Alexa," I asked, "what are African American girls?" I don't know what I expected to hear, but it certainly wasn't the answer I got. Alexa replied, "According to Georgetown.edu, African American girls are the fastest growing portion of the juvenile justice system." When I asked Alexa "what are African American boys," it replied, "According to Edweek.org, the majority of the boys are African American, and many are struggling readers/learners." Alexa just told the child I was pretending to be that African American kids are either "bad" or having trouble with learning.

For a Black child, these answers are devastating at a deeply personal level. For other kids, they plant seeds of, or perpetuate, harmful stereotypes that feed racism. This would be terrible enough in technology aimed at adults. But it's much worse when aimed directly at children by a company claiming it can help with homework. It's not a stretch to imagine that kids today are seeking out all sorts of information on race, gender, sexuality, religion, and more. How will Google and Alexa-sourced information shape children's self-perception and worldviews? Amazon is ramping up its push to get kids even more hooked on Alexa. The company has introduced a new feature claiming to help them learn to read.[23]

Of course, search engines aren't the only platforms where algorithms have been found to encourage and inculcate racism. Social media sites are also culpable. Take Meta, which also owns Instagram and has sometimes been lauded for promoting social justice movements like Me Too and Black Lives Matter.[24] Yet the company has also been under fire for its long history of encouraging hate speech and perpetuating the growth of white supremacy groups.[25] Within a week of George Floyd's murder, a video claiming that his murder was faked reached 1.3 million Facebook users—mostly in groups run by avowedly white supremacists.[26]

To understand how the racism promoted by social networks and other popular tech platforms is linked to commercialism, we need to remember that algorithms governing what content we see and don't see are created by people who, in addition to having their own biases, often work for huge conglomerates whose primary priority is to generate profits for their stockholders. For ad-driven social networks like TikTok, Instagram, Facebook, and YouTube, profits depend on how much companies are willing to spend on advertising—and that depends on how successful the site is at (a) grabbing our attention and (b) holding it for as long as possible. After all, the more of us who use an ad-supported site and the longer we remain, the more exposed we are to its embedded advertising and the more lucrative the site becomes.

So, what keeps us scrolling? Powerful incentives can include outrage, comfort, alleviation of loneliness, and reinforcement of our core beliefs. Meanwhile, every click we make on social media, every "like" or emoji we post, becomes fodder for influencing what news, information, and advertising gets sent to our screens. What we see next is based not on truth or social justice or what's best for humanity, but on what's likely to capture and hold our attention.

Of course, the major social media sites are purportedly just for teens and adults, so you might think their biases or record of choosing profit over truth or accuracy would have no effect on younger children. Yet, tweens and even younger children have been using sites like YouTube, Snapchat, TikTok, Instagram, and Facebook for years.[27] Just as adults do, kids turn to social media for information about the world.[28] And also as adults do, kids use social media to represent themselves to the world, including posting curated selfies and other pictures. But the "self" they present isn't necessarily really what they really look like. That Instagram and Snapchat, among others, provide tools enabling users to "beautify" themselves is problematic on so many levels. For young girls, already self-conscious about their appearance, these tools send the message that their looks aren't acceptable. And, unfortunately, the tools include the capacity to lighten skin tones, exacerbating the race-related problem of colorism, or bias against dark skin.[29] Even without the ability to digitally lighten skin, society's deep and problematic color biases are baked into social media. For instance, in 2021 it was proven that Twitter's algorithm* for cropping photos favored lighter-skinned people over people with darker skin.[30]

What makes the racism embedded in the new technologies so pernicious is that, while companies generate profit by creating narratives about real-world people, places, things, and events, they prioritize content from lucrative sources and give us information we want or "like," not that is necessarily most accurate or truthful. But corporations profiting from the pretend world of children's play—the manufacturers of mass-marketed toys and "old" entertainment technologies like television and film—also have a long history of creating narratives that promote racism.†

In August 2019, I found myself shopping for toys in a major department store in Tijuana, Mexico. It was the last day of a week spent volunteering with the Border Rights Project of Al Otro Lado, an immigrant-rights organization. I spent most of my time at a community center taking care of children while their parents met with lawyers and advocates in hopes of somehow successfully navigating the increasingly draconian asylum process for entry into the United

---

*Twitter's algorithm was also found to prefer younger and slimmer people.
† I am focusing this discussion on toys, tech, and other electronic media. Of course, children's books are also a source of racial stereotypes, but I am not discussing them here. For more information, see Roy Preiswerk, *The Slant of the Pen: Racism in Children's Books* (Geneva: World Council of Churches, 1981).

States. One of my fellow volunteers posted a request on social media for donations to buy toys. As a result of the generosity of her friends and family, I got to go toy shopping with two young almost-lawyers from University of Michigan who were extremely kind about, and apparently even grateful for, my could-be-annoyingly-stringent requirements for what I was willing to buy. We got two small but sturdy sets of tables and chairs, lots of art supplies, and a bunch of nonelectronic and, mostly, unbranded toys.

While the law students searched out games for older kids, I headed for the doll aisle. Since I was in a Latin American country, I assumed I could easily find at least a couple of dolls with brown skin. I was wrong. Every doll on the shelves was white. After my initial dismay, I looked through them again, thinking that I would at least be able to find dolls with brown or black hair. Wrong again. Most were blond, the exception being a couple of bald baby dolls. Not one looked even remotely like the children I passed walking to and from the community center, or like those I met there.

The stories we tell children and the toys we give them are so much more than mere entertainment. They are important components of what social scientists call "material culture," defined as the aggregate of any society's predominant, tangible, human-made creations. A society's material culture simultaneously reflects and influences the values, norms, preferences, and taboos of that society. Stories and toys represent a significant component of the material culture belonging to childhood, and they profoundly influence how children make sense of the world around them, including how they view and experience themselves and others.[31]

Today, the bulk of children's stories (including TV programs, apps, video games, social networks, and books) and toys (often linked to the aforementioned media) are mass-produced, promoted, and distributed by a small number of conglomerates and consumed by millions, if not billions, of children. Deeply embedded in these narratives for kids are lessons about who is powerful or powerless, beautiful or ugly, heroic or cowardly, good or bad, smart or stupid, strong or weak—and even who is visible or invisible. Inadvertently or not, children's stories and toys are rife with societal biases—including about race and ethnicity.*

To understand the power the tools of commercial culture wield over children's attitudes about race and ethnicity, it's important to remember that the notion that children "don't see color" or are oblivious to race and ethnicity is false. Kids pick up on racial differences in the first few years of life.[32] And without thoughtful adult intervention, children can adopt—for life—the prevailing societal attitudes and status of racial and ethnic groups, including their own.[33]

In the United States and other first world countries, children's leisure time is so inundated with tech that it's gotten easy to dismiss the power and influence of hands-on toys. But my experience of being unable to find brown-skinned dolls for the Central and South American children with whom I was working isn't trivial. Because dolls take human forms, they inherently transmit obvious messages about the people they do and don't represent. In her powerfully painful novella *The Bluest Eye*, Toni Morrison describes in detail the layers of rage and longing the little girls in her book feel when confronted with society's ideal of beauty and goodness embodied in the white, blond dolls they receive at Christmas and the white, blond darlings celebrated on screens.

I was reminded about *The Bluest Eye* by a friend and colleague who teaches college students about the intersection of race, gender, and commercial culture and identifies as Afro-Latina. During our conversation, she described her joy when, in 1980, she discovered the brown-skinned Hispanic Barbie amid a sea of white-skinned Barbies at her local Kmart. "Even as a young child I knew that Barbies were supposed to be beautiful. I thought she was

---

*For the purposes of this discussion, I am focusing on race and ethnicity. Of course, they also shape attitudes about gender, sexual orientation, abilities, and more.

so glamourous—and her skin was the exact color of mine," she explained. "I remember looking at her and thinking that maybe I could be beautiful and glamorous someday. I don't think that thought had ever occurred to me before. I couldn't wait to bring that doll to day care! I'd never owned a Barbie before. My white friends were always bringing their white Barbies in. I couldn't wait for them to see that I had a beautiful Barbie of my own that looked like she could be a member of my family."

Her story is so multilayered. She, I, and so many of our colleagues have, for decades, written and spoken out about the negative influence Barbie and other impossibly bodied fashion dolls can have on girls' perceptions of what it means to be female. That Mattel added a Hispanic Barbie to its collection almost twenty years after the brand was introduced doesn't cancel the sexualized and materialistic messages long embedded in the brand. And, honestly, I wondered briefly if it truly *is* a good thing to feel included in a doll line that, in the world of children's toys, has long represented the epitome of male fantasies about women. But—and this is important—however we might feel about commercial culture in general and Barbie in particular, it's crucial to honor the validity of my friend's delight as a little girl in her Hispanic Barbie. It's unfair to hold children responsible for longings generated by corporate marketing that targets them or their peers.

That said, we can and should hold corporations responsible for inculcating and perpetuating harmful biases, whether they are about paradigms of femininity or masculinity, about race and ethnicity, or both. It's particularly telling that when Hispanic Barbie was introduced, the doll could be found primarily at stores serving Latinx populations.[34] While the buyers for that store in Tijuana likely assumed that brown children would want white dolls, it's also likely that the powers that be at Mattel assumed that white children wouldn't want a brown Barbie. Perhaps someone decided that marketing only white Barbies to white children made financial sense, but it also sends white kids the false and damaging message that white skin is the default and dominant human condition. In addition, Hispanic Barbie and the Black Barbie produced in 1980 were both made from the exact same mold as the white Barbies—essentially erasing the existence of features associated with Indigenous and African ancestors of Latinx children or the African heritage reflected in the faces of Black children. It took twenty-eight years for Mattel to launch a line of Black Barbies with facial features approximating those of Black girls and women.[35]

Decades of pressure from activist groups have wrought significant changes in how commercial entities portray various racial and ethnic groups in mass-marketed products. But as one Black father told me, "It's better than it used to be, but even today, it's challenging to find toys that really do look like my kids or their relatives." My colleague Amrita Jain, who has long worked with young children in Delhi, tells me that the same is true in India. "Of course, most of the children in India can't even afford to buy dolls," she said. "But if you walk into a toy store, most of the dolls are white. And even if there are a few dolls dressed in clothes identified with Indian culture, they have light skin and European features." She credits activists in India who are making a difference there—but it's still a problem. "The women starring as romantic leads in Bollywood films, even today, are still mostly fair," she explained. "Whiteness is still associated with beauty here."

The harms of omission long embedded in the cultural tools mass marketed to children falsely elevate whiteness to the default, or preferable, race and erase everyone else. Children are validated when versions of themselves are represented in the toys and media they consume and violated when they never see themselves or are only marginally represented. But they are also harmed when they see depictions of their race or ethnicity that are distorted, exaggerated, or just plain false. In other words, absence isn't the only race-related harm perpetrated by the toy and entertainment industries.

The United States, among others, has a painful history of simultaneously populating children's mass-produced toys and entertainment primarily with white characters while creating and mass marketing degrading stereotypes of people with African, Asian, and Indigenous ancestry. One of the most powerful arguments for how children's toys, and by inference other forms of mass-marketed children's entertainment, simultaneously reflect, reinforce, and ingrain racial stereotypes, is presented in a book by two archeologists. In *Historical Racialized Toys in the United States* (an unfortunately dry title), Christopher Barton and Kyle Somerville explore mass-produced toys from the abolition of slavery in 1865 to the 1930s. According to Barton and Somerville, "Material culture depicting racialized and racist imagery of non-Whites was not the product of individual craft, but mass-produced items that were manufactured, advertised, and consumed by the millions."[36]

The toys they discuss were produced by factories during that period in the postslavery history of the United States characterized by laws specifically designed to legalize segregation and squelch the rights of freed slaves, as well as by lynching and other violent acts mainly against Blacks, but also against other races and ethnicities.[37] It includes, to name just a few, the founding and revival of the white supremacy organization the Ku Klux Klan (1866, 1915),[38] the Tulsa Race Massacre (1921),[39] and *Plessy v. Ferguson*, the Supreme Court decision that codified segregation (1882).[40] Given that history, it's tragic—but not a surprise—that the vast majority of these 172 racialized toys (83 percent) depict demeaning images of Black people. The other 17 percent mostly include demeaning images of Native Americans and Asians. In addition, and to my surprise, two of the toys were racialized depictions of Irish immigrants. I discovered that one tool Britain used to justify colonizing Ireland was to characterize its population as both subhuman and nonwhite, and that characterization migrated across the Atlantic to the United States.[41]

These toys were sold to millions of children through the mass-marketing tools of the day, including catalogs mailed to homes from companies like Sears and Montgomery Ward.[42] Whether intentionally or not, the toys contributed to indoctrinating white children to the legitimacy of racial oppression and the superiority of their whiteness by diminishing or negating the humanity of the Black people whom they allegedly portrayed. Given the racial segregation imposed at the time, for many white children, these toys may have represented the only experience they had of Black people.

The production and sale of overtly racist toys may have abated in the 1930s, but the following decades saw racial stereotypes featured in popular film and television programming aimed at children and families. Popular Disney films routinely featured demeaning stereotypes. These include, but are not limited to, *Peter Pan* (Native Americans), *Dumbo* (Blacks), *Lady and the Tramp*, (Asians), and *Aladdin* (Arabs). In 2019, when Disney's streaming service Disney+ debuted, the company added a ridiculously weak disclaimer to some of its films, refusing to acknowledge the reality of its racist portrayals. "This program is presented as originally created. It may contain outdated cultural depictions." It's the disclaimer equivalent of the faux apology "I'm sorry you feel that way."

In 2020, in the wake of the worldwide protests decrying racism, Disney joined other corporations racing to proclaim their commitment to diversity and altered their disclaimer.

> This program includes negative depictions and/or mistreatment of people or cultures. These stereotypes were wrong then and are wrong now. Rather than remove this content, we want to acknowledge its harmful impact, learn from it and spark conversation to create a more inclusive future together.[43]

When it comes to young children, these disclaimers are useless. Many kids watching these movies can't even read, and many watch these films alone. Nor can we count on parents to be comfortable talking about race and stereotypes with young children.[44]

Meanwhile, the damage done by exposure to these stereotyped images may well be in instilling unconscious rather than conscious bias. In some instances, these may be a white child's first experience of different races and a child of color's first encounter with a negative portrayal of their race or ethnicity. The fact that these films are still available means that millions of young children watch them each year, they are lucrative for Disney, and the racial biases in them are perpetuated.

Of course, the United States is by no means the only country whose entertainment industry has been imbued with racism. When *Consuming Kids* was published in Brazil in 2006, I was invited to give some talks there. On my first day in Rio de Janeiro, my hosts arranged a meeting with a producer of educational children's television programming who showed me some wonderful animation made by children from the favelas, the city's most economically impoverished areas of Rio. But then I was shown an animated film produced by her company for children that featured a small Black character whose face embodied some of the most degrading stereotypes of Black men. His protruding lips and buglike eyes resembled many of the forest animals in the film, including anteaters. His behavior also replicated demeaning stereotypes of Black males—alternating stupidity with cowardliness and a grinning subservience.

It was my first day in a strange country, and while I knew I had to say something, I didn't trust myself to speak up at the time. Instead, giving myself time to think, I raised my concerns in an email, expressing my worries about children viewing such a demeaning stereotype featured in the film. In reply, she justified the film by explaining that it depicted a traditional Brazilian folk legend. The legend is about a spirit who lives in forests and protects the flora and fauna from poachers and anyone doing the forest harm. It's a story with a positive message, especially today when the Brazilian rain forest is so threatened. But retellings of the story don't have to include characters that demean and stereotype Black men. My colleagues in Brazil tell me that as in the United States and India, activists are making progress in forcing companies to address racism. It's true that my encounter with this particular film was over fifteen years ago, but it was posted on YouTube in 2017 and remains on the company's website today.

I understand the importance of transmitting a nation's culture through folktales. I also understand, and share, the nostalgia associated with the Disney films that the parents and grandparents of today viewed as kids and want to share with a new generation. But when these movies are vehicles for transmitting harmful stereotypes, we need to rethink their value and consider the consequences of continuing to show them to young children. I suppose it may be possible with older kids who can engage in abstract thinking to use the films as springboards for talking about racism. But for young children, I believe that the power of the images will eclipse any possible discussions.

Without thoughtful interventions, we know that young children are likely to adopt the prevailing societal attitudes about race and ethnicity. They develop biases based on who and what they encounter in their families and communities. Given the amount of time children spend with toys and electronic media, it makes sense that these also serve to influence their attitudes and beliefs about themselves and other people.

With that in mind, it's important to remember that a corporate mandate to prioritize profit over anything else means that addressing implicit or explicit bias in their products is not automatically a priority. If the changes they make have a deleterious impact on profit, we can't count on corporations to maintain those practices. In addition, since the adults producing and marketing children's toys and media were also once young children absorbing societal biases, it's likely that whatever they create will perpetuate those biases—unless they make a conscious, ongoing, and concerted effort to ensure that they don't.

I've written elsewhere about the ways that companies targeting kids exploit children's innate desire to belong by sending the false message that consumption is a path to acceptance.[45] It's a particularly pernicious message for any child hurt by being excluded by peers. And it's an especially cruel message for kids excluded for any immutable characteristic, like race or ethnicity. My friend's long-term experience of Hispanic Barbie is telling. "Once I had the doll, the white girls liked playing Barbie with me. But I never had the impression that they liked *me* very much. I knew that even with a Barbie of my own I was not accepted by those girls in any deep sense. I was never invited to their homes for sleepovers. They never expressed any interest in me as a person. And they made hurtful comments about my hair."

We can't consume our way out of racism. On the surface, owning Hispanic Barbie changed my friend's daily interactions at day care for the better. But purchasing a particular doll, toy, device, or any product could never alter the kind of deeply ingrained biases she encountered there. As it does today, that would have taken conscious, ongoing efforts at every level of society, including a commitment from her day care's staff to work at effecting change in themselves, as well as the children and families they served.

9

Branded Learning

There needs to be one place in society where children feel that their needs come first—not their future as consumers. In American society today, schools are the only option.
—MARION NESTLE, author of *Food Politics: How the Food Industry Influences Nutrition and Health*

The other day I received an email from the mom of a seventh grader in Montgomery County, Maryland, asking if I knew about the Junior Achievement Finance Park. I didn't, so we arranged a phone call. "I chaperoned my son's class on a field trip to one," she tells me. "They're like a pretend city block with fake storefronts run by national and local brands and featuring their logos. Actually, Finance Parks are all over the country, but the one my son's class went to had Capital One, Geico, Wells Fargo, Volkswagen, and some local businesses and utility companies. And when you go inside the storefronts, they contain promotional pamphlets.

"Kids are given iPads," she continues, "and assigned a pretend persona, home, income, family, and yearly income. They use the iPads to figure out how to get a mortgage, or buy a car, or get insurance. It's supposed to help them learn budgeting. But if they're staring at an iPad all the time, I don't understand why we've taken time out of their school day to bring them here. What's the point?"

It's a good question. The trip is a capstone for a six-week in-school program that's supposed to teach kids how to manage money. Trips to Finance Park are touted as a way for students to "explore and learn about different budget categories, create a budget, make purchasing decisions, and pay for their expenses."[1] But one look at the pitch Junior Achievement makes to potential corporate "partnerships" and it's clear that for participating businesses the point is branding—another opportunity to instill lifetime brand loyalty in future adults.

For $25,000 annually, companies at the park get the following perks:

- storefront signage prominently displayed for approximately ten thousand students, teachers, and corporate and community leaders each year of sponsorship
- the opportunity to fully brand the exterior and interior of the storefront, including (if desired) repainting to brand-appropriate colors and installing company furniture or flatscreens
- the opportunity to distribute informational materials within the storefront to approximately seven thousand middle and high school students
- their company logo displayed on iPads throughout the simulation[2]

In addition, corporate sponsors get to "customize the software content on iPads during Research and Shopping phases of Simulation, including 10–12 careers that exist within the sponsoring organization's business."[3]

Have you visited your children's school recently? They might be learning about nature from teaching materials promoting Disney properties like *Pirates of the Caribbean*, *The Jungle Book*, or *Finding Dory*.[4] Perhaps they're learning about energy production and consumption through the lens of multinational oil and gas companies like Shell and BP.[5] Maybe they've attended an assembly about addiction presented by the e-cigarette company Juul.[6]

Their inspiration for reading could be coming from the lure of free pizza from your local Pizza Hut.[7] Or maybe you and your kids are being pressured to attend "McTeachers' Nights," when teachers at their school go to work for a night at a local McDonald's franchise.[8]

Children's in-school access to technology might be brought to them by Big Tech companies like Google or Amazon, whose business models thrive on collecting and monetizing children's personal information, or companies like Apple, that are intent on nurturing brand loyalty for their hardware. It's certainly not surprising that all sorts of companies try assiduously to get their products into schools. From a marketing perspective, schools are an ideal venue for targeting children.

In fact, advertising in schools may be even more effective than the marketing children encounter elsewhere. Not only are students a captive audience, but school has special significance for them. Whatever feelings children have about school—whether it is a positive or a negative force in their lives—there is an expectation that what they are taught there is good for them, in the same way that eating vegetables is good for them. At the very least, children believe that adults connected with schools believe that what happens there is good for kids—and that aura of "goodness" extends to any product schools endorse or advertise.

School also carries a special weight in the eyes of parents, who are likely to see it as essential to their children's future success. They are likely to view teachers as knowledgeable about education and as having their children's best interests at heart. Anything implicitly or explicitly recommended by teachers or administrators carries that veneer of "goodness," which renders both kids and parents especially susceptible to in-school advertising.

Marketing to kids in school isn't new. In the United States, corporations have been targeting students since the inception of public schools. In 1929, the National Education Association published a report warning teachers about accepting "free" corporate classroom materials. The association argued that corporate handouts should only be used in classrooms if they were essential to children's learning.[9] Over the next decade, battle lines were drawn. The burgeoning consumer protection movement was on one side and manufacturers of consumer goods, along with the advertising industry, were on the other.[10] For companies like Heinz, General Motors, and Hershey, creating supplemental teaching materials was a way to promote their products to future consumers. The advertising industry, however, had a broader goal. They wanted to "persuade students to accept advertising and corporate values as part of their educational experience."[11] By tainting consumer activists as anti-American and by co-opting the more moderate consumer groups, business industry groups were able to discourage efforts to keep teaching materials free of commercial interests.[12] By the end of that decade, one in five corporations was sponsoring classroom materials.[13]

In-school advertising began escalating in earnest in the 1990s, but the groundwork for its escalation was laid in the 1980s, during Ronald Reagan's presidency. In 1983, the Reagan administration's Department of Education issued a report on the state of schools in the United States. Among other suggestions, *A Nation at Risk* urged corporations to get involved in schools.[14] At the same time, the federal government began cutting back on money for state programs and privatizing public services, including schools, was gaining traction.[15]

In fact, federal funding has never accounted for the bulk of school funding, most of which comes from states and local cities, towns, and counties.[16] It's an inherently unfair system for funding public education, which favors wealthy communities. Schools predominately serving children from low-income families and children of color receive less state and local funding than other schools do.[17] According to a survey from the Center for American Progress, roughly 4.5 million children from low-income families attend schools that receive about $1,200 less per student than wealthier schools even in their same district.[18]

As public schools struggle financially, corporations dangle all sorts of what the industry calls *partnerships* as panaceas for lack of funds. For corporations competing to sell products to children, any school district experiencing a budget crunch is a gift. Under the guise of bringing much needed cash, supplies, equipment, or services to struggling schools, companies get to advertise by acquiring naming rights, branding sports uniforms, sponsoring assemblies, and more.[19] It's not surprising that financially strapped schools turn to corporate marketers for help. But subjecting students to marketing almost never comes close to replenishing diminishing school budgets.[20]

While rarely contributing significantly to school coffers, marketing in schools does benefit corporations. Advertisers have long waxed rhapsodic about the benefits of targeting students in classrooms, on school busses, in cafeterias, and in gyms. As one enthusiast for in-school advertising proclaimed over twenty years ago, "Marketers have realized that all roads eventually lead to schools."[21] Another explained the appeal this way: "The advertiser gets kids who cannot go to the bathroom, cannot change the station, who cannot listen to their mother yell in the background, and who cannot be playing Nintendo."[22] That's why, way back in 1995, Consumers Union titled their report on school commercialism *Captive Kids*.[23]

Even if incorporating advertising did significantly increase school budgets, it's clear that marketing in schools contributes to all sorts of ill effects in students. Given the links between junk food advertising and children's unhealthy eating, for instance,[24] it's unconscionable that food giants like the Coca-Cola Company, Pepsico, and McDonald's still routinely advertise in schools.[25] But the consequences for children's learning extend beyond the potential harms of a particular product.

Alex Molnar, a research professor at Colorado State University, has tracked corporate influences in schools for decades. He writes that commercializing education serves corporate interests not just by advertising products and services but also by providing corporations with a "podium to disseminate corporate ideas about topics important to their interests."[26] Targeting a captive audience of students allows corporations to "deliver a broader ideological message promoting consumption as the primary source of wellbeing and happiness.[27] And advertising in schools seems to be an effective means of delivering that message.[28]

What you think about school commercialization might depend on how you view the fundamental purpose of public education. My own beliefs align with those of John Dewey, the early twentieth-century philosopher and educator who advocated strongly for schools that nurture and prioritize skills and attributes essential to a democratic populace—critical thinking, curiosity, creativity, kindness, and social responsibility to name a few.[29]

Dewey's philosophy of education was in direct contrast with that of his contemporaries, including psychologist Edward Bernays, the first to use principles of psychology to make advertising more effective,* and Edward Filene, the department store magnate. Bernays legitimized advertising as education in the service of "the conscious and intelligent manipulation of the organized habits and opinions of the masses is an important element in democratic society." He argued that "those who manipulate this unseen mechanism of society constitute an invisible government which is the true ruling power of our country."[30] Filene believed that the true purpose of public education (he called it "educating the masses") was to create a compliant workforce to better mass-produce goods and unquestioning consumers to buy whatever the factories produced.[31] Neither was interested in educating the public to think critically or independently. While their disagreement with Dewey

---

*Bernays called it *propaganda* or *public relations*, but basically he was talking about what we now call advertising or marketing.

occurred more than a century ago, it is central to current disagreements about whether and how much corporations should influence how and what children learn in schools.[32]

It's easy, for instance, to see how marketing in schools undermines Dewey's vision for education designed to encourage democratic citizenship and promotes the vision of businessmen like Filene and Bernays of training students to be an efficient workforce and devoted consumers. Schools exchanging advertising space for funding from corporations like Nike, Coca-Cola, or Pepsico are likely to be less willing to encourage students to critically evaluate the labor practices or environmental impact of manufacturing sneakers or the effect of soda and fast-food marketing on children's health.[33] Regardless of what students are taught overtly in their classrooms, merely incorporating advertising into a school setting covertly blurs the boundaries between for-profit and nonprofit, civil and commercial, as well as fact and hype. In doing so, commercialized schools—wittingly or unwittingly—condition students to be unquestioning targets for advertising and to embrace unthinking consumption as a normal component of daily life.

I believe that schools should be free of all advertising, but currently I'm especially concerned about the ways that commercializing learning interferes with helping children acquire the skills and attributes essential to preserving and perpetuating a well-functioning democracy. These include, but aren't limited to, affording students access to factual information about the world, providing them with tools for thinking critically, nurturing freedom of expression, encouraging a love of learning, and helping them learn to address complex democratic conundrums, such as balancing individual rights with those of a larger community.

That's why I find corporate-sponsored teaching materials—both online and offline—so worrisome. Because they are often slickly produced, require no up-front cash outlay, and can bypass school boards and be sent directly to teachers, they may appear to be a godsend in cash-strapped classrooms. Nevertheless, while

"Sponsored Educational Materials," or SEMS, may not cost money, they aren't truly free. Funded and created by companies or entire industries whose profits depend on fostering particular points of view, SEMS often come at the cost of providing students with fact-based, balanced information.

During my tenure at Fairplay, we were contacted by the organization Rethinking Schools. Scholastic, the mammoth youth publishing conglomerate, was distributing a unit for fourth graders called *The United States of Energy*, purportedly designed to teach fourth graders about different types of energy-producing materials. The problem? It was paid for by the American Coal Foundation, whose mission is to advance "the power, the promise and the pride of America's coal industry."[34]

Fairplay and Rethinking Schools, along with Friends of the Earth and other environmental groups, launched an ultimately successful campaign to convince Scholastic to stop distributing the materials.[35] Our efforts were covered by the *New York Times*.

> The Scholastic materials say that coal is produced in half of the 50 states, that America has 27 percent of the world's coal resources, and that it is the source of half the electricity produced in the nation, with about 600 coal-powered plants operating around the clock to provide electricity. What they do not mention, are the negative effects of mining and burning coal: the removal of Appalachian mountaintops; the release of sulfur dioxide, mercury and arsenic; the toxic wastes; the mining accidents; the lung disease.[36]

I'm proud of our success. Scholastic not only dropped the coal industry materials but also committed to setting more stringent limits on their corporate-sponsored teaching materials. But clearly, more work needs to be done. Today, often in the name of STEM (Science, Technology, Engineering, and Mathematics) education, the fossil fuel industry continues to distribute classroom materials on energy sources and the environment. These materials are designed, however, to promote the industry's vested interests and deny, or understate, the potential harm the fossil fuel industry causes. Students can learn about fracking from Shell Oil Company.[37] They can watch an online video produced by BP (British Petroleum)—remember the Gulf of Mexico oil spill? On the one hand, the video is about how humans can have a negative effect on the environment. On the other hand, it offers only nature preserves and animal crossing signs as solutions, omitting mention of conservation, alternative energy sources, or global warming.[38]

What's happening to schools in Oklahoma, where oil represents a significant chunk of the economy, is even more worrisome. As of 2017, fourteen thousand teachers were using a kindergarten through twelfth grade science curriculum created by the Oklahoma Energy Resources Board (OERB), which is funded by Oklahoma oil and gas companies.[39] Like the BP materials, those produced by the OERB don't include any downsides of oil extraction and refining. And what's worse is that instead of being exposed in only one grade to science through the lens of fossil fuel companies, kids taught by those fourteen thousand teachers in Oklahoma may go all through school without ever encountering an honest and fact-based assessment of the risks and benefits of fossil fuel consumption.

One of the more concerning efforts by a fossil fuel conglomerate to create teaching materials that promote their financial interests is that of Koch Industries, headed by billionaire Charles Koch. Koch Industries ranks among the worst corporate polluters in the United States.[40] So it's not surprising that the conglomerate has been linked to efforts to ensure that school science materials deny the facts of climate change, including the human effects on environmental degradation. In 2017, the Koch-funded Heartland Institute sent free copies of their climate-change–denying book *Why Scientists Disagree About Global Warming* to 25,000 science teachers around the United States, with the intent of sending one to every science teacher in the country.[41]

Koch Industries is also responsible for creating and distributing teaching materials for high school history, civics, government, and current events classes through the Bill of Rights Institute, founded by Charles Koch and funded in part by the Charles Koch Foundation.[42] As I reviewed Bill of Rights Institute materials,[43] I couldn't help but notice that they highlight freedom, individual rights, individual virtues, and the benefits of limited government. These are important concepts for children to learn and are fundamental components of American democracy. What I couldn't find, however, are the equally important counterbalances to these concepts that are also essential to a successful democracy—the benefits of a government whose laws ensure public education, protect the rights of the less powerful, mitigate social and financial inequality, and regulate corporate behavior that harms the environment and the wellbeing of employees. It's not surprising that these concepts are absent. Educating students who will soon be voters to focus primarily on freedom, individual rights, personal responsibility, and limited government lays the foundation for an antigovernment, antiregulatory view of what's best for the United States. Such an unbalanced presentation of democracy benefits Koch Industries by enabling huge profits unencumbered by laws and regulations that mandate responsible labor and environmental practices.

Of course, the fossil fuel industry isn't alone. Lots of other companies from other industries also create SEMS designed to encourage brand loyalty and ignore or minimize the potential harms of their products and practices. One way for companies to defuse public criticism is to develop teaching materials that appear to counter the problems they're accused of creating. From 2014 to 2016, for instance, McDonald's distributed a school-based nutrition program extolling the virtues of fast food. The program ended only when public criticism from parents, teachers, and activists forced the company to shut it down.[44]

Credit card companies like Discover and Visa distribute materials purporting to teach financial literacy to kids of all grades.

Discover's materials are geared for middle school through high school. Visa's materials span Pre-K through twelfth grade and on through college.

Discover's curriculum *Pathways to Financial Success* claims it will "empower middle and high school students to take control of their financial futures."[45] Pretending to be a seventh grader, I worked my way through a unit called "Making Everyday Purchases." I'm taught that using a credit card is a great way to build up good credit. I'm taught that credit cards are better protection against theft than debit cards. I'm taught that credit cards charge interest if I don't pay off my balance each month.[46] Here's what I'm not taught: That the interest rates charged by credit card companies are often exorbitant—and that's why they actually *want* me to pay only my minimum balance each month. That's how these companies make money. That companies charge huge late fees if you miss, or are late with, a payment. That Discover usually charges $27 for a missed payment and that the late fees are added on to ever-accruing interest from past purchases and that fees can increase with additional missed payments.[47]

Demoting myself from middle school to pre-K, I tackle Visa's unit for the youngest grades. Given that it's for kids between the ages of four and seven, I'm relieved that it doesn't try to sell me on credit cards. It does, however, encourage behaviors and values that are more likely to benefit to Visa than young children. Take lesson one, "What Costs Money," where I learn that I "need" money to buy snacks and toys. That's true, of course. But it neglects to tell me that I need money to buy nutritious food, shelter, medical care, or any of life's necessities. Nor does it tell me that paying for necessities must take financial priority over snacks and toys. In fact, what's missing entirely is one of the most basic and crucial lessons about managing money: how to differentiate between wants and needs.

That it's in the best interest of young children to understand this difference is not esoteric knowledge. A quick search on the web results in advice from experts in education, early childhood, finance, parenting, and at least one in accruing wealth—Warren Buffett.[48]

It's easy, however, to understand why it's missing from Visa's take on financial literacy lessons for young children. Educating kids to distinguish between needs and wants is a step toward developing lifelong habits of spending less impulsively, accruing less debt, and saving more money, all of which threaten Visa's profits.

I move on to lesson two, "Spending Plans."[49] I'm asked to cut out and paste together three envelopes and told to label each either "spending," "saving," or "gifts." I learn that the purpose of saving is to "have more money to spend later" and that if I don't have "enough money to buy something now" I can "save more money and then buy it."[50] What I don't learn, because it's not in the lesson, is what I'm ultimately supposed to do with the money I put in the "gifts" envelope.

I suspect that the cursory inclusion of gifts in the materials is a nod to the growing belief among early childhood educators, family finance advisers, and social justice advocates that along with teaching kids that money is for spending and saving, it's also for sharing as well. Experts who—unlike the executives at Visa—have no financial stake in what young children learn about money stress the importance of helping young children learn to divide whatever cash they accumulate into three containers labeled either "share," "save," or "spend" and to talk with kids about how to use the cash that accumulates,[51] including where they might want to donate or "share" their money. Young children might not understand the meaning of *social responsibility*, *inequality*, or *philanthropy*. Nevertheless, when I taught preschool, the children in my class understood and were passionate about the concept and the importance of sharing.

Visa's teacher's guide for "Creating a Spending Plan" claims that this unit introduces children to the concept of dividing their money into the categories of share, save, and spend. Yet, aside from asking kids to make a "gift" envelope, the materials focus on spending or saving "so you'll have more to spend." From the point of view of what's profitable for credit card companies, it makes sense they teach kids that the importance of money is its centrality to consumption and to satisfying their personal desires. From the point of view of what's best for children, however, that's an incomplete message and therefore a potentially destructive one. Any message about the critical importance of generosity to a humane society is missing entirely from these materials. Kids need to learn about using money not just to benefit oneself but to help make the world a better and fairer place. Instead, like the science materials produced by energy companies, the financial literacy materials produced by Visa and Discover communicate a philosophy about money management imbued with values that promote individual consumption.

That brings me to a different kind of consumption. I was interested to discover that Impossible Foods, a purveyor of plant-based meat substitutes, is carving out a presence in schools. It began with a pilot program in several states where Impossible Burgers, each sporting a little flag with the company's trademark, were sold in school cafeterias. I was particularly interested in the following quote from Pat Brown, the company's CEO.

> Schools not only play a role in shaping children's dietary patterns, they play an important role in providing early education about climate change and its root causes. We are thrilled to be partnering with K–12 school districts across the country to lower barriers to access our plant-based meat for this change-making generation.[52]

I definitely believe that schools should teach about the root causes of global warming, and I understand that massive herds of factory-farmed cattle are an environmental hazard. But, upon reading this quote, I was puzzled about how merely serving Impossible Burgers in schools would educate kids about climate change. That's when it dawned on me that Impossible Foods might be developing

SEMs. It turns out they are. And that's a problem. The company's methods for producing fake meat products that taste like meat are controversial. Environmentalists have expressed concerns about its use of genetically modified soy,[53] while food safety experts have raised questions about its health effects.[54] When I asked Impossible's marketing director whether the materials are going to address the controversies around genetically modified soy, she sidestepped my question.

I'm all for schools teaching the science behind climate change. For health reasons and for the environment, I come down on the side of encouraging people to eat a lot less beef. But, while I might agree with the stated mission of these proposed materials, I wonder whether Impossible Foods, with its financial stake in converting students to customers, can create teaching materials that encourage kids to think critically about the causes of and solutions to global warming.

The rise of digital technology has transformed the way corporations distribute teaching materials in schools. But the creators and owners of those platforms also create SEMS. In the summer of 2017, Big Tech joined the host of other giant industries creating curricula to address the very problems they cause. That's when Google launched *Be Internet Awesome*, school-based lessons and activities designed to teach the company's version of "digital citizenship and safety" to third through seventh graders.[55]

The timing wasn't coincidental. After coasting along on consumers' love of all things digital, tech companies had begun facing unprecedented public criticism. The previous few years saw a significant uptick in concern about the industry's culpability in promoting disinformation, scams, assaults on privacy, various digital addictions, cyberbullying, children's exposure to pornography, and more.

Google wasn't the only Big Tech company receiving bad press (Meta, for instance, was under fire for enabling Russian interference in the 2016 U.S. presidential election). But in 2015, Google was the subject of three high-profile Federal Trade Commission complaints from advocacy groups. One complaint focused on YouTube's practice of unfair and deceptive marketing to children, including ads disguised as entertainment videos for brands like McDonald's and Barbie, unclear boundaries between ads and content, and unboxing videos that appear to be user generated but are funded by toy companies.[56]

Another complaint documented inappropriate content on YouTube Kids. According to a *Wall Street Journal* blog post, advocacy groups found "explicit sexual language in cartoons; jokes about pedophilia and drug use; activities such as juggling knives, tasting battery acid, and making a noose; and adult discussions about family violence, pornography, and child suicide."[57] A third complaint exposed Google's practice of using the company's Apps for Education, a suite of apps offered to schools for free, to violate students' privacy. *Education Week* reported that the complaint was filed because Google was "violating the voluntary Student Privacy Pledge by tracking the millions of students who use its Apps for Education tool suite when they venture to other Google applications such as Maps and YouTube, then using the information to create behavioral profiles."[58]

Be Internet Awesome consists of five units and four snazzy video games artfully designed in Googlesque colors and graphics. Kids are taught to be careful about what they share, to create safe passwords, to refrain from online bullying and to report it when they see it happening, to recognize when they're being scammed, and to talk with an adult if they encounter anything online that makes them uncomfortable. These messages are fine as far as they go, but—as with teaching materials from Visa, Discover, and Koch Industries—I'm troubled by what's missing.

Take privacy, for instance, which Google addresses only in the limited frame of personal responsibility. The message? Any and all violations of your privacy online are your fault. If you are vigilant about what you post and with whom you share it, your privacy will be safe online. Google neglects to mention the modern-day truism about commercial media: You're not the customer. You're the product. This phrase was originally coined about television.[59] Given, however, the tech industry's capacity to track online behavior, email and IP addresses, location, and more, it's more important than ever that kids learn that if something is free online—and often even if it we pay for it— then you, in the form of your personal information, are what's being sold. Further, the materials don't share the crucial information that tech companies like Google amass their gigantic profits by routinely harvesting vast amounts of personal information from users. Google aggregates the data they collect and creates user profiles that they sell to marketers who exploit users through personalized advertising. In addition, as one of the few critiques of the program points out, "the program ignores risks incurred through organizational data breaches and portrays trust in organizations like Google as a sound Internet safety strategy."[60] Nor do the materials mention the very real risk that any personal information shared with a company is vulnerable to being hacked.

In addition to teaching half-truths about online privacy, the materials omit other important information for children. They don't talk about tech addictions of various kinds or even hint that kids might benefit from spending less time on their devices and more time outside, getting exercise, reading, playing alone and with friends, or engaging in countless other activities proven to be both beneficial and satisfying for children.

Google's introduction to *Be Internet Awesome* proudly notes that "the International Society for Technology in Education (ISTE) completed an *independent* [my italics] audit of 'Be Internet Awesome,' recognizing the program as a resource that prepares young learners to meet the 2019 ISTE Standards for Students. ISTE has awarded Be Internet Awesome with the Seal of Alignment for Readiness."[61]

What the materials fail to mention is that Google is one of ISTE's many corporate members and is consistently a gold sponsor of its annual conference,[62] to the tune of $95,000.[63] Given that ISTE relies on tech companies for funding, the possibility of a truly objective audit of any tech materials is highly unlikely. That Google's a member of ISTE and major sponsor of its annual conference renders the term "independent" meaningless. Google also belongs to and funds conferences for another endorser, the Family Online Safety Institute (FOSI), whose membership consists of tech and communications companies as well as related trade associations.[64] ISTE and FOSI weren't the only Google-funded organizations to publicly endorse *Be Internet Awesome* when it launched. So did the National PTA, which has Google as a national sponsor.[65]

Deciding how we, as a society, choose the information that children should be taught in school is a herculean and complex task that goes well beyond the scope of this book. Every curriculum developer, textbook author, or creator of any kind of teaching materials has values and a point of view that influence the information they choose to provide to children and how to present it. As the late historian and educator Howard Zinn said, "In a world where justice is maldistributed, historically and now, there is no such thing as a 'neutral' or 'representative' recapitulation of facts."[66] That means the lessons we choose to teach students aren't neutral—and those lessons matter.[67]

My point in writing about SEMs is not to consider whether progressive or conservative points of view should prevail in school curricula or which facts should or should not be included. It is a discussion about the harms of letting for-profit agendas influence what we teach students. Conflict is inevitable between corporations, whose priority is to generate and protect profits, and schools, whose priority is to generate and protect the education and wellbeing of children. This is especially true for industries whose profits depend on customers remaining largely unaware of any negative consequences resulting from the use or manufacturing of whatever their companies produce.

Fossil fuel industries profit from widespread ignorance of

the human causes of global warming and pollution. Understanding those causes, and what can be done about them, is essential to children, who will certainly bear the brunt of environmental degradation. Profits for fast- and junk-food companies depend on unthinking consumption of unhealthy food. Children, at risk for life-threatening conditions of obesity and type 2 diabetes, need honest, fact-based information about nutrition. Profits for credit card companies increase when customers consistently carry an interest-accruing balance. Kids growing up in an increasingly cashless world and at risk for credit card debt as young adults need to understand the temptations and consequences of spending money they don't have. Big Tech companies, profiting from selling personal information to advertisers, need users who don't think much about privacy policies. Children, as the objects of surveillance that enables powerfully personalized web advertising, benefit from in-depth understanding of how and when their personal information is usurped and exploited.

What kids are taught directly in classrooms and what they absorb from their school environment reflect and shape our societal values. In a democracy, disagreements about the content taught in schools are inevitable. It is not, however, inevitable that schools become vehicles for promoting corporate profits and, in doing so, deprive kids of crucial opportunities to develop their capacity for critical thinking.

10

Big Tech Goes to School

I wonder how many times my kids' teachers use the word "Google" in any given class. "Assignments" and "book reports" are now "Google docs." Doing research is now "googling." When you are both a noun and a verb, you know you've cornered the market!
—Lisa Cline,
co-founder of the Student Data Privacy Project

I am chatting with a reporter who writes for a parenting website. During our conversation, he mentions his two young sons, so I turn the tables and ask him about his family's experience with marketing to kids. "Do you know about Prodigy?" he asks. I don't. "My son's teacher recommended it for my eight-year-old son. She told us it was a free online math game and that we should get it so my son can practice at home. My wife and I limit our kids' screen time, but since his teacher suggested it and it was about math, we figured it would be fine. But the same day he started playing at home he began asking us to buy him the premium version. Sure, you can get the game for free, but every time my son plays, he sees ads telling him why it's better to get the premium version, which is definitely not free."

The frequency and effectiveness of corporate marketing in

schools has historically been tied to the ever-increasing power and sophistication of available media technologies. In the 1950s, when I was a kid, we sometimes got to watch "educational" films in school. By 1954, over 3.5 million students watched about sixty thousand showings of films from the National Association of Manufacturers. The American Gas Association taught us how to make a sandwich—tuna rarebit cooked in a double boiler over a "simmer flame" on a gas stove, then served on bread toasted in a gas oven.[1] Shell Oil celebrated the importance and safety of fossil fuels.[2] And even as General Electric dumped the toxin polychlorinated biphenyl in the Hudson River, the company positioned itself as a foe of water pollution by distributing the film *Clean Water* in schools around the country.[3]

As entertainment and communication technologies evolved, corporations found increasingly powerful ways to advertise in schools. From 1989 through 2018,[4] millions of kids in thousands of junior high and high schools around the United States were required to watch episodes of Channel One News daily, a twelve-minute cable news program, which included two minutes of commercials, created exclusively for schools.[5]

Channel One is defunct, but it's important for two reasons. It was the first time kids in school were exposed daily to tech-enabled advertising, and its sustained nature allowed researchers to evaluate the quality of its content and the impact of its advertising. The educational value of news piped to students in those twelve minutes of their school day was questionable.[6] Those two minutes of advertising, however, were remarkably effective. A study out of the University of Missouri found that students not only remembered the ads, some even reported dreaming about the commercials.[7] In another study, ads on Channel One were found to promote preferences for the products advertised. Even more concerning, they also appeared to influence students to adopt the underlying materialistic messages embedded in all commercial advertising.[8] Students required to watch Channel One daily were more likely to agree with statements such as "money is everything; people with more money are happier than people with less money; designer labels make a difference; a nice car is more important than school; people with more money are happier than people with less money."*

These days, those corporate-sponsored films and even Channel One seem quaint compared with the invasive and persuasive marketing strategies employed by today's digitally driven edtech industry.

One huge challenge to writing about the business of educational technology is that "edtech" is an umbrella term for a wide range of products and services that can be divided roughly into three categories. For instance, systems like Google's Workspace for Education are primarily marketed as tools for organizing and streamlining classroom management. These systems can provide ways for students to work on assignments and submit them digitally. They may also allow teachers to—among other things—monitor students' progress, communicate with parents, and store students' personal information, including grades, attendance, and more. As advocates have long pointed out, one big detriment to students is the potential violation of their privacy if tech companies have access to their personal information. Such information is a valuable commodity that could be sold to advertisers, for instance, or shared with police or other entities.[9]

Another category of the edtech industry is the practice of distributing hardware like individual iPads and Chromebooks to students as young as kindergartners. A great deal has been written about the problems associated with what media theorist Criscillia Benford calls "hardware dumping," the notion that the mere act

---

*The researchers compared students in a school using Channel One with students at a school that didn't. They made sure the schools served communities with similar racial and ethnic makeup, socioeconomic status, and access to commercial television. These similarities enhance the probability that watching or not watching Channel One influenced the students' responses.

of handing kids their own iPad or Chromebooks, for instance, will improve their classroom performance.[10] It doesn't.

The Organization for Economic Co-Operation and Development studied millions of high school students around the world and found that those who used computers heavily at school "do a lot worse in most learning outcomes, even after accounting for social background and student demographics."[11] A recent global survey found that when it comes to learning reading, math, and science, the best results come when students are in classes where teachers alone use tech as an aid in their teaching.[12] In the United States, research suggests that on average, kids using tablets in "all or most of all" of their classes have reading scores the equivalent of a full grade lower on assessments than do kids who never used tablets in their classroom. In some states, kids using tablets most of the time scored even lower.[13]

Clearly, handing out scads of devices without a clear, evidence-based plan for when, how, and how much to use them isn't an effective means of educating students. The primary beneficiaries of hardware dumping are the companies like Apple and Google who make the devices and their operating systems. Google, for instance, charges a management fee of $30 per Chromebook,[14] but that's only a partial benefit. The biggest boon to Google is the likely potential of lifetime brand loyalty from millions of students using the company's operating system and products for years before they graduate. As Mike Fisher, an edtech analyst at the market research firm Futuresource Consulting told the *New York Times*, "If you get someone on your operating system early, then you get that loyalty early, and potentially for life."[15]

Edtech also includes an ever-increasing number of digital games and "personalized learning tools" like Prodigy, the math app that so annoyed the reporter who told me about it.[16] These apps and "learning games" purport to teach kids everything from math to reading, yet there's little independent research documenting their effectiveness or the wisdom of using them to replace or diminish the duration or frequency of interactions between students and teachers.[17]

A statement released by Fairplay in 2020 and signed by over one hundred organizations and individuals (including myself) with expertise in education, health, and early childhood, put it this way: "The value of quality, teacher-driven instruction is well supported by research. There is no credible research supporting industry claims that online, personalized learning programs improve academic outcomes. Test scores do not rise. Dropout rates do not fall. Graduation rates do not improve."[18]

Prodigy is merely one of many such products targeting children today, but it's worth describing in depth because it is increasingly popular. The game is produced by a Canadian company that recently changed its name from Prodigy Game to Prodigy Education and was named one of Canada's fastest-growing companies in 2019.[19] A Prodigy Education press release claims that the app is currently used around the world by 50 million students and 1.5 million teachers.[20] According to the same press release, the company's mission is to "help every student in the world *love* learning."[21] What appears to drive schools to recommend it, and teachers to let kids play it in their classrooms, is that the game is positioned as a "fun" way to help children learn math.

I spent many hours playing Prodigy over the course of writing this chapter, logging in as different children of different ages by lying multiple times about my grade level, name, and country. The game combines elements of successful children's media properties like Harry Potter (wizards, wands, and spells!), Pokémon (choose your own loveable little creature to be your very own pet!), and Fortnite (Battle! Zap others to win! Play with friends! Dance!). In addition, Prodigy incorporates many of the persuasive design techniques proven to capture and hold children's attention—intermittent rewards, customizable avatars, levels, virtual thingies to acquire, and more.

Prodigy consists of battles between wizards and with various other creatures. Along the way, players spend time picking and personalizing an avatar (by selecting hair, skin, and eye colors from a variety of options), selecting magic weapons, learning crazy dance steps, and more. It's kind of like a low-rent version of Fortnite.

But wait! What about the math? It's true that every time players engage in a skirmish, they win by solving a math problem. Yet math isn't really integrated into the game. One of the first things I noticed about the game when I first started playing in 2019 is that the math problems could easily be lifted out and replaced with questions about social studies, science, or any classroom-based subject. I suspected that Prodigy Education was eventually going to expand its offerings beyond math. That proved to be true. In October 2021, Prodigy began promoting a 2022 version called Prodigy English that is supposed to teach "English Language Arts."[22]

Since I'm not an expert on teaching math to kids, I asked Rheta Rubenstein, professor emerita of mathematics education at the University of Michigan–Dearborn, to spend some time playing Prodigy. Her response: "Not only is math totally extrinsic to the game, there's nothing in it to help students actually learn math (understand, find relationships, use reasoning, solve novel problems, defend an argument, and so on)."[23] But even if Prodigy did a good job of teaching math, its business model is still problematic.

What's particularly troubling about Prodigy being used in public schools is that it's what tech marketers call a "freemium." It's marketed as absolutely free while earning profits by subjecting children to repeated, in-game, seductive ads and inducements, promising that kids will have ever so much more fun and move up ever so much faster if they get their parents to upgrade to a paid "premium" version. The pressure to convince "my parents" to let me upgrade to the premium version of Prodigy began the instant I chose my avatar. Sometimes it was a pop-up ad urging me to get a "free 7-day membership today." Sometimes I was treated to an animated video featuring tinkling music and a chirpy narrator urging me to go premium so I could "enjoy monthly membership boxes with loads of stuff, leveling up 50% faster . . . cool gear, extra wheel spins, evolving all of your pets, and awesome cloud to ride and more. Ask your parent or guardian to help you today."[24] In a complaint to the Federal Trade Commission, my colleagues at Fairplay noted that "over a 19-minute period we saw 16 unique advertisements for membership, as well as opportunities to see ads via shopping and social play, and only four math problems. That's four ads for each opportunity to concentrate on math."[25]

As I continued to play, the pressure to upgrade mounted. Not only would I get more cool stuff if I nagged my parents to join, but I would "level up" faster. And leveling up faster became more appealing every time I was offered magic spells to help me in combat but was told that I wasn't at a high enough level to obtain them. Sometimes I was offered a jar chock full of virtual stars and told that if I didn't become a member in a few days, the stars would disappear. And so on. When I first began playing Prodigy, what the children I was pretending to be were never told, but what "my parents" would find out once I successfully convinced them to upgrade, is that a premium account would cost them anywhere from $4.99 to $8.95 per month. That's $59.88 to $107.40 annually—per child.[26] But as of August 2021, Prodigy raised the cost of a premium membership (now called a Level Up membership) to $74.95 or $119.40 each year, depending on whether parents buy a monthly or annual subscription.[27] The company has also added a new tier, Prodigy Ultimate, which costs from $99.95 to $179.40 annually. The company claims that Prodigy Ultimate is their "most comprehensive plan to help maximize fun AND learning."

In addition to leaning on those millions of kids to nag their parents for an upgrade, Prodigy also pressures parents directly. On its parents' portal, Prodigy claims—with no visible evidence—that "Premium Members advance beyond their grade level in a matter of months by tackling more math challenges to level up and receive special rewards."[28]

As proof of the game's effectiveness at teaching math, the

website prominently features the following quote from research conducted by Johns Hopkins University: "Increased use of Prodigy was significantly correlated with increased achievement gains for students on a standardized mathematics assessment."[29] A look at the actual study, however, shows that "a student would need to complete roughly 888 questions in the game in order to achieve a one-point gain in their standardized assessment score."[30] That's a lot of questions, taking up a lot of time. If students average just a minute answering each question, they're spending more than 14.5 hours with Prodigy—and that doesn't include the time spent getting new pets and equipment, choosing spells, and engaging in other activities that have nothing to do with math. Students could spend the equivalent of a whole day and night playing Prodigy to get that one-point increase.

Prodigy's pitch to parents also claims that kids using the app will love learning math.[31] Since I could find no research proving that Prodigy is more effective at helping kids learn math than any other technique is, or that kids playing prodigy learn to love math, I called the company's support line to ask whether they had any research. They don't. I'm not surprised. There are lots of ways of helping kids learn to enjoy math—or at least experience its usefulness and its connection to their lives and activities. Depending on children's ages, these include building with blocks, conducting surveys, measuring themselves, each other, and things around them, constructing models, and more. In these activities, working with numbers, measures, diagrams, scales, and number concepts is essential to the experience.

In Prodigy, despite that solving math problems is the way to win battles, math is something to be gotten through in order to have fun in the rest of the game. In fact, one message children could take from Prodigy is that math is like medicine disguised in applesauce or pudding—so distasteful that the only palatable way to consume it is if it's immersed in something much more appealing. Supposing, however, playing Prodigy did make children love math? What if it was the best way to reinforce the math lessons taught in school? Its business model, based solely on manipulating children to pester their parents to pay for an upgrade, still renders it unsuitable for use in schools. As one teacher commented in a review: "I DISLIKE IMMENSELY that there [are] constant popups about getting a membership. I do realize you need to make money however many of my students can not afford a membership and it makes them feel bad when they are constantly assaulted by that."[32]

The problem goes way beyond the hurt feelings or embarrassment kids might feel when they're exposed to continual pop-up reminders to buy an upgrade their parents may not be able to afford. But Prodigy also foments envy and shame to get kids to convince their parents to buy a membership. The site functions as a modified social network, which means that players can see other players. They can see who has the shiny badge identifying them as paid members with more privileges—and who doesn't. They can see how members get to earn more points and acquire more virtual stuff. And in a particularly cruel twist, they can see how members can float around on clouds while nonmembers have to trudge along in the dirt.[33] Prodigy creates what amounts to different tiers of students in the classroom. As Josh Golin puts it, "What makes that worse is that everyone can see who has a membership and is getting the cool stuff and who isn't. I think the inequality is one of the clearest examples of prioritizing profit over what's good for kids."[34]

Prodigy exemplifies the escalating presence of commercially driven edtech apps and games in schools. Of course during months of the pandemic, school was completely virtual for millions of children, which was a boon for programs like Prodigy. But even before the COVID-related school closings, edtech was a booming business. In 2019, the edtech market in the United States alone brought in $28.3 billion.[35] That same year, a Gallup poll showed that 89 percent of third to twelfth graders used edtech products in school at least a few days a week.[36] The same poll showed that half of students were happy with how much tech they use and that

42 percent would like to use it more often. Of course, kids love all sorts of things that may or may not be good for them—or that may or may not be great tools for teaching and learning. According to Gallup, school personnel, including teachers, principals, and administrators, were also enthusiastic about digital teaching tools—although significantly more principals and administrators than teachers were fully supportive of their increased use in the classroom.[37]

In addition, for most of the school personnel surveyed, their enthusiasm for tech didn't stem from a knowledge of supporting evidence. According to the report, fewer than three in ten teachers say there is a lot of information available about the effectiveness of the digital learning tools they use. Only about half say they have some information about them, while almost one-fourth say they have little or no information at all.[38] Meanwhile, an assessment of the effectiveness of edtech in the *MIT Technology Review* suggests that school personnel should be tempering their fervor.[39]

Marketers are notoriously good at identifying societal trends or movements and co-opting the words used to describe them to attract buyers for whatever they're selling. Take the word "green," which was adopted by environmentalists in the early 1970s as shorthand for relating to or supporting the natural world. As the environmental movement gained traction, marketing experts began warning corporations that they'd better win "the loyalty of the growing legions of green consumers."[40] Green morphed into a common marketing buzzword employed even by fossil fuel companies and airlines, which are notorious for their harmful impact on the environment. Green was such a misused descriptor that in 1986 an environmental scientist named Jay Westerveld coined the term *greenwashing*: the practice of companies advertising their products and practices as environmentally beneficial when they verifiably are not.[41]

I found myself thinking a lot about greenwashing as I researched edtech products and kept encountering the term "personalized learning" in their marketing. It's currently a tag line used to market edtech programs like Prodigy that are designed for kids to use on their own without input from teachers. The term is used to maximize the use of digital technologies in children's learning. In doing so, it minimizes, and even dismisses, the central importance of teachers to the learning process.

In fact, research shows that teachers are essential to effective "personalized," or "personal," learning—whether kids are using edtech materials or not.[42] As Alfie Kohn, author of *Punished by Rewards* and other books about education, said in *Psychology Today*: "[true personal learning requires] the presence of a caring teacher who knows each child well" *and* "works with each child to create projects of intellectual discovery that reflect his or her unique needs and interests."[43]

In reality, "personalized" or "personal" learning predates edtech by decades. Like the word *green*, it's been corrupted by the marketing industry's practice of exploiting a social movement—in this instance, the theories and practices of progressive education—and using it to sell products that have little to do with and are antithetical to the original meaning of the term.

The meaning of "personalized learning" is rooted in research and practice pointing to the following conclusions: Children have an innate drive to learn, and how they learn best varies from child to child. Kids are not passive, empty vessels waiting to be filled with facts but rather active, innately curious explorers.

Two concepts connected with progressive education's version of personalized learning are particularly intriguing to me. One is "constructing knowledge" and the other is "making meaning." The phrase "constructing knowledge" evokes a vision of kids actively participating in learning and that what they've learned serves as a foundation on which to build their understanding of new information they encounter. Meanwhile, the term "making meaning" describes the human drive to understand, make sense of, and relate to whatever they encounter. In education, making meaning suggests that real, usable learning occurs when children grasp a concept so deeply that they can actively apply what they learn in one context to challenges that arise in another context.[44]

If you want to experience children constructing knowledge and making meaning, you might want to hang out for a while with newly verbal young children as they encounter the world. They often narrate thought processes that older children have learned to keep internal. When my daughter was a toddler, for instance, she encountered a black olive for the very first time. After studying it a while, she looked up and announced, "This is not a grape!" She'd encountered something new (the olive) and, on her own, felt compelled to understand what it was. She searched through her twenty-two or so months of life experience for clues to make sense of it until she found one. While she did not know what it was (an olive), she at least knew what it was not (a grape)!

My understanding of personalized learning also comes from teaching in a play-based preschool. And my daughter's experience attending a play-based preschool reinforced my belief in its value. In each instance, kids had access to materials like books, art supplies, blocks, sand, water, dress-up clothes, and special projects that, for the most part, they could explore in their own time, depending on their interests. Children's involvement with the materials was driven by their interests and inclinations, but teachers were always available to join in, advise, supervise, stand back, observe, or help kids reflect on their experience.

A 2019 report from the National Education Policy Center at the University of Colorado is a sweeping condemnation of the edtech version of personalized learning. It found "questionable educational assumptions embedded in influential programs, self-interested advocacy by the technology industry, serious threats to student privacy, and a lack of research support."[45]

Thinking about the "questionable educational assumptions" embedded in Prodigy leads me directly to the popular phenomenon of gamification, or gamified learning, which applies some of the more addictive features of video games to subjects taught in school. These can include badges, levels, digital prizes, competition, and variable rewards.

Gamified edtech products are a lucrative business these days. Globally, game-based learning is expected to garner $29.7 billion in 2026, up from $11 billion in 2021.[46] The rationale proponents often give for gamifying education is that kids like video games and sustain their interest in them for hours at a time.[47] It makes sense, the reasoning goes, to transfer the gaming features that keep kids glued to screens to classroom teaching and learning. And, since these products are games, and games connote play, it also makes marketing sense to link these products to the robust evidence that play is the foundation of intellectual exploration and crucial life-enriching abilities such as problem-solving, reasoning, literacy, social skills, creativity, and self-regulation.[48]

One obvious difference is that when products lean heavily on external motivations like competition and virtual prizes, they teach kids to dismiss the value of experience and they promote the value of acquisition. In contrast, the kind of play that facilitates children's learning, growth, and development is its own reward. It's a deeply satisfying experience in and of itself. Opportunities for actual play-based learning help kids learn that the world is an intriguing place and that exploring it and figuring things out are both interesting and valuable in and of themselves.

Edtech companies try to link their products to play by touting them as "fun." For instance, an ad for Prodigy on YouTube proclaims that it will "make learning math fun!"[49] In fact, like "personalized learning," "fun" is a big component of marketing for gamified edtech products. Versions of the trope "make learning fun" turn up repeatedly in edtech marketing, with headlines like "Gamification: Make Learning Fun with Genially"[50] or "Khan Academy: Making Learning Fun."[51] Tag line variations on "make learning fun" may generate a lot of sales, but they have problematic implications for children's learning *and* their capacity to have fun.

The *Oxford English Dictionary* defines fun as "light-hearted pleasure, enjoyment, or amusement; boisterous joviality or merrymaking; entertainment."[52] I believe that fun is crucial for children's wellbeing—and my own. I initiated my Audrey Duck chats with kids at the beginning of the pandemic because I wanted to do my part to provide opportunities for unadulterated silliness in the midst of a very stressful time. What I didn't anticipate, however, was that the fun these chats generated was an ongoing source of delight, energy, and hope that helped me get through my own pandemic-related stress.

Fun is good, but it's a false promise to imply either that learning should always be "light-hearted" or "amusement" or "entertaining" or that the opposite of fun is boredom. Sometimes learning is fun, but it can also be satisfying and engrossing even when it's hard work. Have you ever witnessed the intensity and drive of a baby learning to walk or a child learning to ride a bike or master any skill they decide to command? I wouldn't describe falling in the interest of learning to ride a bike as fun, but kids get back on and try again because learning to ride matters to them. Learning is rewarding when it has meaning, not just when it's "merrymaking" or "entertainment."

Even as edtech marketing pitches like "make learning fun" distorts the process of learning, the content of the heavily marketed edtech apps and games I looked at promote the same false narrative I describe in chapter 5. Having fun in these games depends on external rewards like applause, cheering, special effects like showers of stars, or virtual prizes. It's a great message for edtech marketers looking to inculcate consumerism in kids, even when they're in school, but it's a lousy life lesson for children.

Of course, it's certainly possible to use digital technologies to promote meaningful learning. Apps that allow children to truly create or "build" do that, such as the process of coding or designing and constructing actual robots. An evidenced-based article in *Psychological Science in the Public Interest* suggests that based on the science of learning, apps can facilitate meaningful learning if the children using them are "*actively* involved ('minds-on'), *engaged* with the learning materials and undistracted by peripheral elements, have *meaningful* experiences that relate to their lives, and *socially interact* with others in high-quality ways around new material, within a context that provides a clear learning goal."[53]

This is fine, as far as it goes, but the article neglects to address the central problem with the for-profit edtech industry, which by definition has to prioritize revenue over the students they claim to serve. Suppose, for instance, an app meets all of those criteria, but—like Prodigy—generates profit as a freemium by pressuring kids to nag their parents to pay? Suppose its revenue comes from ads or in-app purchases? Suppose, in collecting data on students with the stated purpose of allowing teachers, parents, and administrators to monitor academic progress, companies sell students' personal information to other companies, who in turn target kids with behavioral advertising or use their information in-house for that purpose?

As the technology for virtual and augmented reality evolves, edtech is only going to get more compelling. Educators are already enthusing about the learning made possible by, for instance, immersing kids in virtual renditions of the prehistoric world.[54] Unfortunately, with notable exceptions like the analyses coming out of the National Education Policy Center or the efforts of advocacy organizations, public discourse on the pros and cons of incorporating edtech tends to omit any description of exactly how these apps generate profit and the ways that these features can harm children. Who will own the virtual worlds students explore? Will these worlds contain embedded advertising? Will students' privacy be protected? Will data be collected on students' activities in virtual worlds? What will happen to that data?

It's understandable that when the pandemic forced schools all over the world to rush headlong into educating children remotely, decisions about edtech were made without much time to think them through. But, under normal circumstances, it's in the best interest of children that we all, including teachers, administrators, and school boards, approach edtech offerings with healthy skepticism.[55] And, like any materials used in schools, edtech programs, platforms, and devices should be free of any features that exploit kids for profit.

11

Is That Hope?

No one can seriously suggest that children are rational consumers who have the same power, information, and freedom that adults are said to have to freely enter into contracts for goods and services. Advertising to children is, then, a kind of immoral war on childhood, waged for the profit of adults who should be childhood's guardians.
—Alex Molnar and Faith Boninger,
"The Commercial Transformation
of America's Schools"

In 2014, novelist Russell Banks was invited to give the Ingersoll Lecture at Harvard Divinity School. Established in 1893, the lectures are based on the theme of immortality. Given that Banks is a self-described atheist, he was an interesting choice. Instead of discussing immortality in the context of religious or spiritual beliefs, he spoke beautifully and poignantly about children as our afterlife. Specifically, he talked about the damage done to children, and therefore to the survival of our species, when they are immersed in a commercialized culture of our own creation.[1]

In an existential sense, given the links between overconsumption and the destruction of the planet, children as afterlife is a starkly powerful metaphor for why it's imperative to protect them from manipulation by marketers. But there are also so many "here and now" reasons as well. In these pages, I've focused primarily on the harms of a commercialized culture on children's relationships, values, and learning. But allowing marketers unfettered access to children is also a well-documented factor in a host of other problems facing children today, including childhood obesity, sexualization, youth violence, and underage alcohol and tobacco consumption. Today, corporate influences on children's entertainment, leisure activities, and education are so pervasive that it's both unfair and unrealistic to expect parents alone to bear the burden of shielding kids from industries spending billions of dollars to capture children's hearts and minds en route to their parents' hard-earned cash.

That said, parents certainly have an important role to play with their own children, and I provide suggestions for things they can do in chapter 12, but the real solution lies in systemic change.

We need to end what amounts to a corporate takeover of childhood. And despite protestations to the contrary, history tells us clearly that we can't rely on industries to stop targeting kids on their own when their profits are based on hooking children.[2]

Regardless of documented links between smoking, heart disease, and various cancers, Big Tobacco continues to target children.[3] So does Big Food. In 2006, in response to growing alarm about childhood obesity and calls to regulate food and beverage marketing to children, the food and beverage industry banded together with pledges to foreswear junk-food marketing to kids.[4] More than fourteen years later, not much has changed—children are still heavily targeted with ads for junk food.[5] And Big Tech is no different.

The roots of any particular social ill often lie in deeper systemic problems. Today's commercial assault on children is tied to government policies that contribute to the escalation of corporate power and the decline of government regulation, diminished support for public schools, parks, and play areas, the rapid evolution of digital technologies, and the consolidation of media ownership. That's why we need to address broader sociopolitical issues that leave children more vulnerable to exploitation by Big Tech and other industries that thrive on marketing to kids.

Campaign finance reform, especially reversing the current practice of allowing corporations to funnel unlimited funds to political campaigns, will curb corporations' undue influence on all sorts of government policies, including those affecting marketing to children. Adequately funding public schools, including prekindergarten and after-school programs, will make it less likely that schools will rely on corporate advertising for support or use SEMS in classrooms. Providing support for and maintenance of safe, accessible outdoor spaces for children to play and spend time in nature could reduce the amount of time children spend indoors on their devices.

In the United States, two types of laws would help to stop tech companies from exploiting children. A national privacy protection law, which we do not have, and adequate laws protecting the rights of children, which we also do not have. In fact, the United States is now the only country that has not ratified the United Nations Convention on the Rights of the Child.[6]

Usually, the first step toward any kind of social change is changing public opinion. When Fairplay began, and until recently, the likelihood of getting any kind of legislation passed was pretty much nonexistent. We met with members of Congress and we filed FTC complaints, but our focus was on building support by highlighting some of the worst examples of marketing to children. Most of our efforts focused on corporate campaigns like the ones I mentioned earlier—from preventing Hasbro from releasing dolls based on a burlesque troupe[7] to stopping Disney from marketing Baby Einstein videos as educational.[8] Among other successes, we closed down a company putting commercialized radio on school buses[9] and stopped McDonald's from advertising on children's report cards.[10] But times have changed. Fairplay and other advocacy groups still organize corporate campaigns, but they are now also actively involved with legislators at both national and state levels who are working to regulate the ways tech companies target children.[11]

For the first time, around the world, there are positive signs for advocates trying to curb some of the more egregious tech-based forms of marketing to children. As a news headline in the *Australian Financial Review* proclaimed, "Finally the World takes on Big Tech."[12] Of course, the phrase "takes on" is key. It's true that governments around the world are taking steps to curb the power of tech conglomerates like Amazon, Google, and Meta, but of course these companies are fighting back.[13]

When it comes to children and tech, perhaps the most hopeful news comes from Britain. In 2020, the British Parliament adopted the Age-Appropriate Design Code, which became law in September 2020 and went into effect a year later. The code mandates that "the best interests of the child should be a primary consideration when you design and develop online services likely to be accessed by a child."[14] What's particularly important is that it defines a child as anyone under eighteen, which protects teenagers as well as younger children.

It's also extremely important that the code doesn't just leave it up to companies to decide what's in the best interest of children. Instead, the code requires a stringent set of default settings for companies with "online services likely to be accessed by a child."[15] These include high settings for limiting data sharing and ensuring privacy. These restrictions are intended, among other things, to prevent companies from using children's behavior and personal information for personalized advertising. They are also intended to prevent companies from sending kids push notifications designed to extend their engagement with particular content. In addition, anything kids post online will automatically be private, which means the content can be seen only by friends.

It's a victory for children that the code requires default settings that clearly prioritize what's best for children over what's best for corporations. But "default" doesn't mean the settings necessarily have to stay that way. Children can choose to override the defaults, which would, for instance, make their posts to social media publicly available. What could have a much broader, and potentially worrisome impact, however, is that companies can also change default settings by demonstrating "a compelling reason to do so, taking into account the best interest of the child." At present, we just don't know how this is going to play out in real time. For instance, we don't know what kind of arguments companies might devise for how changing the defaults will be in the best interests of children. Nor do we know how the British government will interpret what "best interest of the child" means or how stringently these rules will be enforced. What we do know, however, is that when it comes to protecting children from exploitation by tech, media, and marketing companies, relying on corporations to self-regulate doesn't work. For that reason alone, Britain's adoption of the Age-Appropriate Design Code is a significant step toward protecting children.

From an international perspective, what's important about the code is that its rules extend beyond British companies to include any business offering any online product that might be used by kids living in the United Kingdom. Making strict privacy settings the default won't eliminate advertising to children, but kids are less likely to be easily manipulated by the advertising they do encounter online because it won't be curated based on the data companies collect about their individual interests and vulnerabilities. Companies compliant with the code, for instance, will no longer be able to identify kids susceptible to smoking or drinking and target them with ads designed to encourage their interest in these and other unhealthy behaviors or share information about children's emotional states with advertisers.

Britain may be the first, but other countries are also beginning to step up.[16] In the United States, the Children's Online Privacy Protection ACT (COPPA) went into effect in 2000. It requires that websites targeted to children under the age of thirteen obtain verifiable parental permission before collecting or using personal information from children or sharing any of that information with third parties. But it's woefully out of date. COPPA was enacted before tech behemoths such as Google, Snapchat, and TikTok honed their capacity to surveil children and to monetize whatever data they collect. As Angela Campbell, professor emerita of Georgetown Law said in her recent testimony before Congress, "When COPPA was adopted in 1998, there was no YouTube, no social media, no smartphones, no smart speakers in children's bedrooms, and no toys connected to the internet."[17]

I have to say that, in my eyes, Angela is an unsung hero. For more than twenty years, she and her students at Georgetown's Institute for Public Representation, working pro bono, enabled nonprofit organizations, including Fairplay, to file numerous complaints urging the FTC to use its legal authority to prevent tech and media companies from exploiting children. In fact, in 1996, the institute filed the very first such complaint against a tech company's website for targeting children with unfair and manipulative marketing. This ultimately led to Congress enacting COPPA. Angela retired from the institute in 2020, but the work she began continues today, and she remains a leading expert on, among other things, laws relating to media and technology as they affect children.

In her 2021 testimony to Congress, Angela pointed out that another long-standing problem is not with COPPA itself but with a lack of will to enforce it stringently enough to make a difference.

> Non-compliance with COPPA is rampant. . . . In fact, in the 21 years that the COPPA Rule has been in effect, the FTC has brought only 34 enforcement actions, mostly against smaller companies. All were settled without litigation by consent decrees. Often, settlements merely required the defendant to comply with the law and file periodic reports with the FTC. When the FTC has assessed civil

penalties, they have been woefully insufficient to incentivize compliance with COPPA.[18]

The hopeful news for children and their advocates is that the landscape seems to be changing. For the first time, several bills that would significantly limit the ways tech companies can exploit children are wending their way through Congress. Whether any of them become law is another question, but just the fact of their existence is a big step forward—not the ultimate step, but one that signals an important societal shift. Public outrage about the ways children are exploited online has grown strong enough that legislators are finally looking for ways to curb it. And, for the first time ever, legislative efforts to restrict Big Tech's exploitation of children have the explicit support of a sitting U.S. president. In his 2022 State of the Union address, President Joe Biden called for strengthening privacy protections online, banning targeted advertising to children, and preventing tech companies from collecting personal information from children.[19] With that in mind, it's worth looking at what passage of these bills would achieve.

It would be best if there were just one bill providing full protection for children, but in aggregate the bills currently being considered are impressive. In the House, the Kids PRIVCY Act mirrors Britain's Age-Appropriate Design Code by stipulating that the wellbeing of children become a primary web-design consideration, extending protections to kids eighteen and under, and applying the restrictions to any website likely used by them.[20] The bill authorizes an outright ban on advertising to children driven by data collected about their behavior and personal information. In addition, for areas not covered explicitly in the bill, it empowers the FTC to decide exactly how ensuring children's wellbeing transfers to web design.[21] Of course, there's no way of knowing what the FTC would decide. Currently, the FTC chair, Lina Kahn, has a history of calling for regulation of huge tech conglomerates.[22] For first time, it's possible that Big Tech would no longer have license to invade the privacy of children and teens in the United States and exploit their personal information for profit.

One bill in the Senate, the Kids Online Safety Act (KOSA), is promising because it was introduced with bipartisan sponsors. It provides for an independent outside audit of sites targeting kids, requires tech companies to consider the wellbeing of children into their design, empowers States Attorney General and the FTC to hold tech companies responsible for actions that harm children, and requires that all default settings for children be set at the most protective.[23]

Another proposed bill, the Kids Internet Design and Safety Act, known as the KIDS Act,[24] addresses unscrupulous tech marketing techniques that target children and teens under the age of sixteen—not just on platforms overtly aimed at kids but also on platforms like Instagram and Snapchat that ignore the fact that they are heavily used by children.[25] If passed, the KIDS Act will protect children and young teens from techniques designed to keep them online indefinitely by banning autoplay, push notifications, endless scrolling, and those streaks on Snapchat that can keep kids up all night. It will also protect them from sneaky, insidious marketing that disguises itself as content by banning influencer marketing, like unboxing videos such as Ryan's ToysReview. The bill addresses content as well as design. It "prohibits websites for kids and teens from amplifying violent, inappropriate, and dangerous content."[26] In addition, it provides funding for commercial-free educational content for children and funding for academic research on the impact of screen-based media and content on children.

States are also challenging Big Tech and the entertainment industry. State legislatures around the country have passed, or are considering, bills to protect children's privacy online and to prevent tech companies from selling personal information collected online. Twenty-nine states and Washington, DC, now limit or prohibit screen time for children under two in day care and other early childhood settings.[27] And the Minnesota state legislature is considering a bill that has bipartisan support that would limit the use of individual screen devices in preschool and kindergarten.[28]

In addition, multiple bills on a state and national level have been introduced that require teaching media literacy in schools.[29] I believe there's value in helping kids learn to decode media messages and identify Big Tech's marketing techniques, but my support comes with a caveat. Media literacy is important, but it's crucial to remember that we cannot and should not rely on children alone to protect themselves from screen-based marketing. Focusing solely on media literacy without limiting corporate practices is kind of like blaming victims for the harms perpetrated on them by others. And given how vulnerable adults are to manipulation by tech and entertainment companies, it's neither fair nor reasonable to expect children to protect themselves.

12

Resistance Parenting: Suggestions for Keeping Big Tech and Big Business at Bay

Navigating technology for children can be overwhelming. The truth is your child doesn't need apps to "keep up," "catch up," or "get ahead." You have the full power to limit, or say "no" to screen time.
—Rachel Franz, Fairplay, *Safe, Smart, and Secure: A Guide to Choosing Tech for Your Preschooler*

There's no question that despite its numerous joys and satisfactions, being a parent is challenging. It's even more challenging when life gets complicated by stressors like financial worries, job pressures, marital discord, illness in the family, and more. We can and should aspire to be the best parents we are capable of being—but we must also let ourselves off the hook for those inevitable times when our reality doesn't meet our aspirations. And one stressor unique to parents today is the beckoning omnipresence of captivating and seductive technologies designed to capture, hold, and monetize our attention and our children's.

The most consistent and insidious way Big Tech and big businesses harm kids is to lure them to digital devices from birth, keep them hooked, and train them to be unthinking and uncritical consumers. In and of themselves, and for better or worse, our phones, tablets, and other devices are powerful tools for influencing human behavior. What makes tech use turn toxic for children and families is rooted in the industry's prevailing profit-generating strategies. These include invasions of privacy, data-driven advertising, infinite scrolling, and other manipulative techniques designed to capture and hold our attention and leave us yearning for more. The problem is not that young children are harmed if we occasionally hand them a cell phone to play with. The problem is that once we start doing that, it's often a struggle for us and our children not to make it the norm.

As I compiled the thoughts and suggestions I offer, I was reminded how much disparities of wealth and circumstance are a factor when considering the impact of tech on children. Many of the recommendations I offer are so much easier to implement in families headed by two adults raising children in safe neighborhoods, with adequate childcare, and with access to green space. The recommendations are likely harder to implement for those working multiple jobs to stay afloat or who are struggling with joblessness or other kinds of stress—or who live in neighborhoods with few safe parks or outdoor play spaces. I also recognize that some kids are more innately challenging to raise, for instance, children who struggle with impulse control more than others their age do. That's why I believe the following: **The tech, toy, and entertainment industries' practice of monetizing childhood is a problem for all of society that can be solved only through social change.** One family alone can't combat huge conglomerates making use of ubiquitous and irresistible technologies, child psychologists, and billions of dollars to exploit children's developmental vulnerabilities. It is by no means a level playing field, and it's particularly hard for parents under stress. That said, social change takes time and families need help now.

The questions and concerns I hear most from parents today are about how much and what sort of tech experiences we should offer our children, and at what age these should begin. To answer these questions thoughtfully requires a basic understanding of how children grow and develop. Toward that end, I turn to my colleague Nancy Carlsson-Paige, professor of child development emerita from Lesley University and co-founder of Defending the Early Years. The six evidence-based principles of child development she identifies, found on the following page, are a useful guide for deciding how young children spend their time, including whether, how, how much, and when they use digital devices.

In combination with a basic understanding of how children grow, learn, and develop, it's also helpful to consider the available research on the effect of tech and media on young children, including their potential benefits and harms. But this is problematic for two reasons. For one, it's unreasonable to expect busy parents to sort through the research themselves. For another, while the body of research available on young children and digital technologies is growing, it's still in its early stages. Basically, we have two choices. One is to acknowledge that research is incomplete, so we tell parents not to worry about young children's time with tech until there's robust research suggesting harm. Another is to employ the precautionary principle. We acknowledge that the research is incomplete, but we urge parents to limit young children's exposure to tech until there is robust research suggesting that it's beneficial, or at least harmless.

When it comes to children's health and wellbeing, the latter makes more sense to me. That's why I think the most useful recommendations based on available research come from the international pediatric public health community.[1]

Around the world, pediatric public health organizations are in near universal agreement that it is best to avoid screen time for babies and toddlers and to limit time for preschoolers to less than an hour a day. When it comes to school-age children, some recommend no more than two hours a day. Others no longer recommend

Six Principles of Child Development to Help Adults Make Decisions About Introducing Tech to Young Children*

1. **Young children live and learn in the context of social relationships.** Children's emotional and social development happens slowly over time, just as their cognitive development does. They develop awareness and skills slowly that grow from their experience interacting with others.
2. **Young children use their whole bodies and all their senses to learn about the world.** A child's whole development, brain development included, is best supported when babies and young children have full-on opportunities to use their whole bodies and senses for activity, play, and social interaction.
3. **Young children learn best and benefit most from direct, firsthand experience in the world of actual relationships and objects.** This is because three-dimensional experiences are wholistic, they involve a child fully—body, mind, and feelings—and this level of engagement is greater than what can be gained from two-dimensional experiences.
4. **Young children are active learners who learn by inventing ideas.** For genuine learning to happen, kids need to construct ideas for themselves, in their own minds. This is the kind of learning that is real and genuine and stays with us.
5. **Young children build inner resilience and coping skills through play.** The more that elements on a screen shape play the less a child's play can come from within. More direction

---

* Adapted with permission from Nancy Carlsson-Paige, "Young Children in the Digital Age: A Guide for Parents," Defending the Early Years, November 2018, dey.org/wp-content/uploads/2018/11/young_children_in_the_digital_age_final_final.pdf.

> from outside means less access to the inner life of imagination and emotion.
>
> The less a child's play comes from within, the harder it will be for them to build inner resilience and coping skills.
>
> 6. **Young children make sense of the world through play.** As adults, we have this ability to use our thoughts and words to process our experiences. But children don't have these tools. The way young children process and make sense of their experiences is through play.

time limits for older kids and adolescents. They do, however, urge parents to work with their kids to develop a family plan for the role tech plays or doesn't play in their home life and to review and revise their plan together as children grow. They urge consistency and emphasize the importance of ensuring that kids spend most of their time engaged in activities known to be beneficial to healthy development, for instance, hands-on active and creative play, face-to-face time with friends and family, reading, and more.

One outlier whose recommendations I find surprising is Britain's Royal College of Paediatrics and Child Health, which opts not to issue time guidelines even for babies and very young children. Instead, their primary recommendation is that "families should negotiate screen time limits with their children based upon the needs of an individual child, the ways in which screens are used and the degree to which use of screens appears to displace (or not) physical and social activities and sleep."[2] This makes sense for older children and is similar to guidelines issued by other health organizations. But it's disappointing that they don't consider the developmental differences among babies, preschoolers, tweens, and teens and recommend no screen time for infants and time limits for younger kids.

What follows are a variety of thoughts, suggestions, and

information about managing tech and commercialized culture at home.* They align with my focus throughout this book on children ranging from newborns through middle childhood, roughly ages ten or eleven. Since I am skeptical of tech solutions to tech-induced problems (the adage about foxes guarding henhouses comes to mind), I am not including or reviewing apps designed to help you or your children manage screen time.

Because the needs of children and their families evolve as kids grow, I begin with suggestions organized according to children's ages and developmental stages. Next, because how *we* use and don't use tech affects our children, I offer suggestions for reducing our own screen time. I also offer some general suggestions to consider as you think about your whole family's device use. Last, but most definitely not least, I offer suggestions for helping kids resist destructive, marketing-induced behaviors and values.

Infants and Toddlers

Babies aren't born yearning for tech. We train them to long for time with tech by consistently handing over smartphones and other devices whenever they need to be soothed, stimulated, or amused. In doing so, we deprive them of opportunities to learn to soothe and amuse themselves, to find stimulation in the world around them, and to generate their own entertainment.

Nor are babies born loving Elmo, the characters from *CoComelon* and *Paw Patrol*, or any other brand icons. Babies do, however, take immense pleasure in what is familiar to them. That's why media companies market character-branded baby paraphernalia, even for newborns. Here's how it works. Infants surrounded by, for instance, Peppa Pig mobiles, clothes, bedding, and toys will start to smile or laugh whenever they see Peppa's image. The hope for companies is that parents and other relatives will interpret those

---

*Thanks to Josh Golin for his help with these suggestions.

expressions of joy as evidence that there is something special about Peppa Pig and will therefore be more likely to buy baby items featuring Peppa's image. And these purchases will add to an infant's attachment to the popular pig, which will encourage even more sales.

- **Avoid starting babies and toddlers on screens.** Except for video chatting with faraway adults who love them, there's nothing necessary or particularly meaningful that babies and toddlers can gain from time with videos, games, or apps. And time with screens takes away from time engaged in activities proven to be beneficial to brain development—actively exploring the world in safe spaces with all their senses, being read to, and bonding with the adults who love them. In addition, the more time babies and toddlers spend in front of a screen, the more time they are likely to spend with screens when they get older.[3]

  So, bearing in mind the aforementioned principles of child development and remembering the habit-forming nature of so much tech content, seriously consider postponing introducing babies and toddlers to digital videos, games, apps, programs—with the exception of video chatting—at least until the age of two and for as long as you *comfortably* can after that. In many ways it will make your life easier. Your kids will have a chance to develop the inner resources to soothe and amuse themselves without depending on screens, and you get to avoid, or at least postpone, the unpleasant experience of conflicts over screen time.

- **Remember that the more a toy can do, the less a child needs to do.** And the less a child does with a toy, the less useful that child's play is to healthy development. Avoid chip-enhanced toys that sing, dance, walk, or talk at the push of a button. Toys that bombard children with sounds take away essential opportunities for babies and toddlers to babble, talk, sing, or make noise of any kind. Toys that move on their own take away opportunities for babies and toddlers to hone their fine motor skills as they push

little cars along or make stuffed animals move. Toys that spring into action at the mere push of a button deprive babies and toddlers of opportunities to develop agency and a sense of competence in the world.
- **Get in the habit of reading to children from infancy.** There's lots of evidence that reading to children beginning when they're babies has multiple benefits. It helps develop the building blocks of language and helps to prepare them to do well in school. Holding a baby on your lap and reading a picture book to them is also a perfect opportunity to cuddle. The combination of reading and cuddling together is an incredible bonding experience for parents and children. And it has the added benefit of leading to a host of positive associations with books. If you choose to use an e-book to read to kids, choose one with no digital enhancements, which have been found to interfere with making sense of the story. As I describe in chapter 7, however, there is evidence that reading analog books instead of e-readers with children—even e-readers without enhancements—is more likely to spur the kinds of conversations that are beneficial to literacy, and more conducive to cuddling![14] Make use of libraries and yard sales to find books. And remember that while reading is essential for children, e-readers are not. Once kids can read on their own, if you want to get them an e-reader and can afford the expense—choose one that only allows for reading. It's harder to concentrate when you're reading on a device that also delivers instant access to games, videos, social media, and other digital distractions.

Preschool and Kindergarten

Early childhood is the easiest time for you to construct family habits, routines, and activities that reflect your values. It's the best chance you'll ever have to shape the choices offered to your children for what to play with, who to play with, and how to play.

The prevailing recommendation for preschoolers from the public health community is no more than an hour a day of screen time. And many families choose to do less than that, or none at all.

- **Remember that there's no evidence that children must start using screen technologies in early childhood to succeed in a digital world.** In fact, the technologies available, like touch screens, are purposely made to be easy to use and are likely to be outdated by the time children today reach adulthood.
- **Begin talking with your children about your choices and the values they're based on.** Keep that conversation going as they grow and develop. When kids start having playdates with other children or enter day care or preschool programs, they most likely will encounter toys and tech that they don't have at home. It's helpful to be able to point out to kids that families, and their values, can be both different and the same as other families.
- **Remember that it's easier to increase children's time with tech than it is to reduce it.** You always have the option to introduce your children to screen-based entertainment, but once you do, it's harder to cut back or take it away all together. That said, a psychologist friend who works with families reminded me that it's important for parents to remember that it is possible to cut back on, or to change, the amount of time children spend with screens. She tells parents, "You can tell kids that you made a mistake and explain why you think the current rules aren't working."
- **Remember that while young children have lots of developmental similarities, they also have different temperaments, interests, strengths, and vulnerabilities.** Pay attention to what your child is like around devices. Is it hard for them to stop using them? Do they nag or have tantrums when you deny their requests to use them? Are they irritable afterward? Is their tech use preventing them from engaging in activities known to be beneficial to them? If any of these are the case, consider cutting back on the

amount of time your children spend with screens or taking a break from them altogether.
- **Be wary of distinctions between "active" and "passive" media.** When apps first appeared, proponents of the new technologies dismissed TV and films as "passive" entertainment because kids can't affect what happens on a screen. They designated social media, or digital apps and games, as "active" because kids can affect screen content primarily by tapping, scrolling, or swiping. But I think this is a misleading distinction. Many interactive apps stifle creative play by offering prepackaged choices, making children's involvement more reactive than active. Meanwhile, linear, story-based television programs or films can evoke deep feelings, encourage us to think, expand our view of the world, introduce us to new words and concepts, and encourage empathy. In other words, engaging with them is not necessarily passive at all. Shows originating from public television, for instance, are often a good option for quality, age-appropriate programs—but they are not necessarily commercial-free. Be prepared to deal with children's requests for licensed toys, clothes, and accessories associated with the programs they watch.
- **Be skeptical of apps labeled "educational" on Google Play or the Apple app store.** App developers get to choose their own categories and are not subject to review. In fact, most apps for young children labeled as educational aren't.[5]
- **Try to pick apps and programs that are age appropriate.**[6] Ideally, we would play games and apps or watch the videos we offer to our children ahead of time. In reality, many—if not most parents—don't have time to watch a whole movie or video or play an app before offering it to their children. That's why organizations such as Children and Media Australia (CMA) and Common Sense Media (CSM), which offer curated reviews of apps, games, films, and television, are a useful family resource. CMA reviews are always free. CSM allows users to read three free reviews a month.

- **Remember that apps advertised as "free" might actually be freemiums.** With freemiums, only a limited version of the app is free, and your child is likely to be hounded to pay for an upgrade and/or make in-app purchases. If possible, let your child know ahead of time whether you're willing to pay for the premium version. And this is another instance where acknowledging the validity of your child's desires is important, even if you refuse their requests to upgrade ("I know you really want to upgrade, but we're not going to do that"). You can also share your feelings about freemiums ("I think it's so unfair for the company to purposely make you feel bad about not owning the premium version").
- **Choose apps that are as commercial-free as possible.** App stores identify whether apps have in-game ads or in-game purchases. Remember, however, that apps featuring licensed characters, even those as beloved Elmo, are de facto ads for any product bearing that character's image. One of the things that's so hard about children's media today is that so much of it is designed to sell kids on related programs, apps, and films as well as brand-licensed toys, clothing, food, and accessories.
- **Choose apps that encourage actual creative activities like freestyle drawing and music making.** Lots of apps claim to encourage creativity but only offer predesigned images or "right and wrong" options for making art.

Middle Childhood / Elementary Grades

While some public health organizations recommend no more than two hours a day of entertainment tech, others urge parents to engage their children in making consistent plans for tech use that work for their particular family. Decisions around tech use get more complicated as children leave early childhood behind. For one thing, they become more susceptible to peer pressure. For another, their social lives are likely to become increasingly tied to tech. In addition, their schoolwork is more likely to require online access.

- **Postpone getting your child a smartphone until at least eighth grade.** When it comes to raising children, smartphones are probably the most pernicious of all tech devices. We carry them everywhere and use them excessively. In doing so, we provide Big Tech with tons of information about who we are, where we go, who our friends are, and more. In return, we are constantly bombarded with personalized advertising designed to hook us where we're most vulnerable. And the more we use them, the more attached to them we become. And remember, Bill Gates delayed getting his kids smartphones until they were fourteen, about the time they were in eighth grade.[7]
- **It's best to start thinking about how you're going to handle the smartphone question when your child is in early elementary school.** Since, as I report in chapter 7, almost 20 percent of eight-year-olds have a smartphone (and there's every reason to believe that number will rise), it's best to start planning how you are going to handle the likelihood that your child is going to ask for one long before you're ready to give one to them.
- **It's easier to postpone getting your child a smartphone if the families of your child's friends are doing the same.** In other words, there is strength in numbers. The organization Wait Until 8th (www.waituntil8th.org) provides a wealth of information and resources to help you bring together the parents of your child's friends to delay getting kids a smart phone until eighth grade.
- **Support children playing with friends in real life.** Remember that digital "sandbox" games that allow kids to play remotely with their friends, like Minecraft, Roblox, and Fortnite, are not a substitute for playing with friends in person. For one thing, they don't satisfy our basic need for human contact. For another, when kids are in the process of creating structures from actual blocks together or running around pretending to fight off space

creatures, their play and their devotion to it is not continually monetized. Yes, the toys or tools they use cost money. But children's actual experience of playing together is not continually monetized. It's not interrupted by advertising. No one tracks their every move in order to target them more effectively or to figure out how to keep them playing longer. Kids playing games like Minecraft and Fortnite are subject to ongoing pressure to purchase digital add-ons like decorations for their avatars or sticker packs to enhance their texts.

- **Remember that social media sites like Instagram and TikTok, which claim to restrict access to kids younger than thirteen, can be problematic even for teens and adults—so it's best to keep younger children off them.** Like so much that tech offers, these sites have the potential to do some good. It's true that they can help marginalized teens find social connections, but they can also be extremely problematic. Since they are especially good at capturing and holding our attention, they lend themselves to overuse—which is linked in teens to depression, lower self-esteem, and negative feeling about their bodies.[8] But there are other problems as well.

  It's hard to sort out truth from fiction on social media, and tech companies still fail to do a good job of helping us do that. And in addition to being primarily platforms for advertising to users, they also foment envy by inviting us to market ourselves to other users. It's hard enough for us to remember that what we and our "friends" post is a curated version of our lives, designed to have the world see us as we wish to be seen, it's even harder for children who tend to believe what they see.

- **Finally, remember that children's versions of these sites, like Facebook Messenger and Instagram's proposed app for kids that is currently on hold, are designed to prime children to use the regular sites—serving the same function that candy cigarettes served Big Tobacco.**

If You Want to Reduce Your Own Time with Tech

Before we can help kids resist the powerful lure of tech-based and commercially driven entertainment and the consumerism they encourage, we need to understand our own vulnerabilities to tech and to commercialism. Children first learn how to navigate the world from parents and primary caregivers. They learn by interacting with us, but also by observing our behavior. So, if we want to keep children's tech use to a minimum and if we want it to be volitional rather than habitual, we need to consciously limit the amount of time we spend with our devices—especially when we're with our kids. Of course, I'm not talking about those times our jobs tie us to our screens. But we need to take an honest look at what hooks us into spending too much time on tech unnecessarily and figure out how to avoid or mitigate those situations.

- **Wear a "dumb" watch.** One brilliant, seductive, and troublesome feature of smartphones is that they combine so many services in one device. And once we take them out for any reason, including to check the time, it's hard not to start scrolling through social media, answering texts, or checking email. Wearing a watch is one easy way to cut down on the number of times you check your phone. And I suggest wearing a "dumb" watch that only tells time. "Smart" watches that replicate features provided by smartphones, such as access to apps, the internet, texting, and more, are just one more device designed to capture our attention and surveil us.
- **Turn off notifications and alerts, at least when we're with children.** Those beeps and buzzes that function as alerts are designed to pull us to our phones and away from whatever and whomever we're engaged with. And not only do they disrupt our focus, they also disrupt the focus of everyone around us. Think long and hard about whether your job or family situation really requires you to respond to texts and emails instantly. One parent, the mom of two

young boys, told me how liberating turning off notifications was for her. She explained, "Now I feel like I'm checking my phone, not that my phone is checking me."

- **Be conscious of when you're primed to pull out your phone unnecessarily and strive to resist the urge.** Just being aware of when you're likely to do it is a step toward gaining more control. As danah boyd, a partner researcher at Microsoft, suggests, if you're with children and you check your phone, tell them why you're doing it, for instance, "I'm expecting a text from your friend's mom to see when we can come visit." She says, "Once you begin saying out loud every time you look at technology, you also realize how much you're looking at technology. And what you're normalizing for your kids."[9]

  Emily Cherkin, who works with families struggling with screen-time issues, advises, "Live your life out loud."[10] Saying out loud your reason for checking your phone is also likely a deterrent to checking it. It might be uncomfortable telling children, "I don't know why I'm doing this, but I feel compelled to," or "I'm checking to see how many 'likes' the picture I posted on Instagram has."

- **If you choose to read on an e-reader, choose one that is solely for reading and without access to other digital distractions.** The pull of having immediate access to all these other platforms can be hard to resist.

- **Find a phone-free thing to do when you're watching kids play on their own or with friends.** Living with constant access to a 24-7 information blitz has made it harder for us to tolerate doing nothing. And, as I describe in chapter 7, our phones can be so absorbing that we get irritable when our children interrupt us. Try to find something engaging to do that will distract you from yearning for tech-induced distraction. If you enjoy making things, you can bring along yarn to knit or crochet, a small pad for sketching, or other portable art materials. Another possibility is to always have something to read that isn't your phone or tablet.

As You Make Decisions for Your Family About Tech and Commercial Culture:

- **Think about your values and priorities and—to the extent that you're able—develop a plan around tech and commercial culture that reflects them.** If you are raising children with a partner, talk together about how tech and commercialization support or undermine the values you want to pass on to your children. Try to come to an agreement about tech use at home and be consistent about how you implement your plan. Children do better when expectations are consistent, and it's much easier to achieve consistency around anything related to raising children when the adults raising them agree on values and priorities.
- **Be skeptical of advice about kids and tech from any blog, expert, or organization accepting funding from tech or media companies.** One of the consequences of corporate funding of nonprofits, such as public health, advocacy, media literacy, educational, or early childhood organizations is that it's potentially dangerous and even catastrophic to offend the hand that feeds you. It's possible that their suggestions may be unbiased, but it's also possible that they are influenced, consciously or unconsciously, by a desire not to alienate their corporate funders. Many nonprofits list their major funders on their website—sometimes they're listed under "partners" or "supporters" or are available in online annual reports.
- **Be wary of convenience.** It's a rare parent who does not at least occasionally opt for convenience over best practices in childrearing. And convenience is one of the major attractions of digital devices, especially for parents pressed for time, resources, and energy. Devices provide a quick, easy, all-in-one way to keep kids happy and occupied and allow us time to get things done, relax, or talk with our partner or friends. But convenience is a slippery slope.

  The problem with opting for convenience in the moment is that it can have unintended consequences in the future. Handing babies or toddlers cell phones or tablets to calm them down or to keep them occupied may solve an immediate problem—but it's crucial to remember how habit forming these devices can be. You are encouraging kids to need a device to amuse or soothe themselves.

It's true that when kids use devices they tend to be quiet and occupied for extended periods of time, which may make things easier in the moment. But postponing your children's device use as long as you comfortably can is likely, in the long run, to make parenting easier. Your kids will have more opportunities to develop strategies for soothing and amusing themselves without depending on devices. You will avoid potential unpleasantness around time limits, exposure to nag-inducing advertising, and endless decisions about what content really is appropriate.

- **Be skeptical of the prevailing myth that children's early device use will make parenting easier.** Josh Golin puts it this way: "Far too often in discussions about young children, a false dichotomy is presented: Either you're on the floor playing with your child or, if you take a break, they're on a screen. In this telling, the needs of the parent who needs time to get other things done, or rest, are at odds with a child's need for offline play."[11] What's missing from the scenario he describes is the fact that it is possible for lots of kids to occupy themselves alone without screens. Try to have toys and materials on hand that fit your child's age and interests. To name just a few examples, these could include art supplies, books, puzzles, things to build with, dolls, and stuffed animals.
- **Carve out consistently tech-free, commercial-free family time.** Make tech-free family meals a priority. Device-free meals encourage family conversations about big and little events of the day. In addition to helping kids get in the habit of separating from devices when they eat, keeping them off during meals sends the message that family time is more important than electronic distractions. Some families choose to extend tech-free time beyond meals. They set aside tech-free days or evenings to play cards or board

games, do puzzles, go for hikes, read aloud, play music, toss a ball, or do whatever they enjoy together.
- **Think about your space and how you want tech arranged in your house.** Try not to make your TV the centerpiece of your home. And if space allows, establish tech-free zones and, regardless, keep devices out of children's bedrooms. Getting enough sleep is essential for children's healthy development and for the wellbeing of adults. Research strongly suggests that screen use before bedtime interferes with sleep.[12] Some families designate a particular time and place where everyone puts their devices away for the night.
- **Spend tech-free time outdoors.** Allow children unmediated time in nature—to play, to connect emotionally and sensuously, to explore, to create, and to wonder. Our goal should be to help them learn to love the natural world and connect deeply to it. Using tech to categorize flora and fauna and to learn facts about them can be fun for children but that's something you can do together when you get home. For young children experiencing nature in the moment, what's most important about encountering a bird, for instance, is appreciating its beauty or the wonder that it can fly and recognizing it as a fellow living creature. Research suggests that children play more creatively in green space[13]—let them build things with sticks, and stones and luxuriate in mud, sand, and water.
- **Provide children with lots of opportunities to generate their own tech-free solutions to boredom.** If your children are already used to turning to tablets or phones when they're bored, this can initially be challenging for all of you. Kids can get cranky and irritable, which is hard on them and on us. If they're used to depending on screens for stimulation, begin the process of weaning them first by planning screen-free activities and then by encouraging them to choose what they'd like to do. It may be bumpy at first, but ultimately opportunities for screen-free play can lead to all sorts of creativity and help children develop the inner resources

to generate their own activities and enjoy their own company and not constantly need outside stimulation.
- **Try to find other parents and communities whose decisions around tech are at least in the same ballpark as yours.** It's harder to buck societal norms on your own. And it's helpful for your children to have at least one friend whose family has similar rules around tech. Talk with other parents about managing tech with children. Share your concerns about it. An increasing number of good books and films are available to schools and communities that are excellent springboards for discussions about tech, kids, and commercialized culture.* Not only do these resources raise awareness, they are also a terrific way to find parents who share your concerns. At least when your children are young, you may be able to develop some shared ground rules for playdates. These rules can include no device use, or time limits for their use, no violent content, and more.
- **Try to get your extended family on board.** It's harder to keep children tech-free if their cousins are constantly playing with apps or if grandparents consistently hand them smartphones to play with. Try talking with relatives to see if you can come to an agreement about how your kids will spend their time during visits. If that doesn't work, you can explain to your children that different people have different values, and while rules about tech may be different at the homes of various relatives, your family has its own ways of managing tech.

Help Kids Resist Destructive Marketing-Induced Behaviors and Values

- Remember that one of the most troubling consequences of our commercialized culture is that it thrives on exacerbating some of the worst of human tendencies—envy, selfishness, unthinking

---

*See page 245 for suggested books and films.

impulsivity, and disregard for the common good. If we want our kids to absorb positive attributes like kindness, generosity, and altruism, then it's important to let them experience our efforts to live lives that reflect those values. Some families involve their kids in taking donations to a food bank or in volunteering at centers that serve meals to people who can't afford food. If you donate money to charities, involve your kids in the process. Some families set aside an evening to support the causes, organizations, and institutions that embody their values. One family I know collects items like shoes or baby books from friends or neighbors to donate to a shelter in their town. Involving kids in acts like these can lead to life-shaping conversations and experiences that will nurture their sense of responsibility to a larger community.

- **Once you start offering kids an allowance, encourage them to manage it thoughtfully.** One way to do this with young children is have them divide their money into three categories designated as "share" (money to donate to charity), "save"(money to put away and add to), and "spend"(money to purchase things they might want immediately).
- **Find ways to help children find meaning in celebration that extends beyond the commercial.** Holidays that involve gift giving can lose their roots in spiritual or cultural traditions, especially when children get deluged with presents. Tell and retell the holiday's origin stories and sing holiday songs together. Prepare food together that is linked to your family and cultural holiday traditions. From early on, include gifts of time and experiences in the presents you choose for your child. Ask relatives to do the same.

Remember that less is often more. Young children can be easily overwhelmed by too much stimulation, including being faced with a ton of presents. Getting too many gifts can lead kids to focus solely on the excitement of acquisition, depriving them of opportunities to pay attention to, and appreciate, what they are receiving.

- **Remember that it's okay to say no to children, but it's important to acknowledge the validity of their desires.** Try to avoid giving in to nagging since doing so will just lead to more. Keep in mind, however, that children's feelings of longing are real, just like ours are. It's both painful and frustrating when our expressed feelings are unrecognized, denied, or diminished by important people in our lives. But validating our children's feelings doesn't mean that we have to accede to their requests. For instance, you can say, "I know you really want this, but I'm not going to buy it for you." And then try to explain your reasons for saying no.

    It's also important to differentiate between expressing a desire for something and nagging, and to help your children do the same. Children need to feel comfortable expressing their desires, so they need to know it's okay to ask for things that they want. And they also need clear and consistent limits to help learn that nagging is not an effective strategy for getting what they want.
- **Before taking your kids shopping, including grocery shopping, let them know ahead of time what you will and won't buy or how much you're willing to spend. Then stick to your decision.** It's easier on us and our children if everyone knows the rules ahead of time.
- **Talk with children about advertising and marketing and share your feelings about them.** While young children can't fully grasp the persuasive intent of advertising, they will start to absorb your attitudes about it. And pointing out less obvious forms of advertising, like product placement or unboxing videos, can help them at least distinguish between ads and other forms of content. Recognize that when your child is exposed to any kind of advertising, digital or otherwise, it is likely to increase their desire for whatever is being sold.

Just as children have similarities and differences, so do families. Some of the suggestions included here will work for you and your children and some won't. Pick and choose. What's important, however, is to recognize that the tech, toy, and entertainment industries have enormous power in children's lives, that their unmitigated influence can be harmful, and that steps can be taken to help your child develop a sturdy sense of self that is not in thrall to commercial interests. What's equally important, and a major reason that I and my colleagues founded Fairplay, is to recognize that other people's children need help as well.

13

Making a Difference for Everybody's Kids

> *People are extremely easy to manipulate. I worry about the ways that technology is, and will be, used for political or commercial purposes. As machines are increasingly able to read our emotions, know what we are doing, and get us to send messages to other people, it will be possible to make almost anything happen. I just think it's very, very scary. What we can do through education and legislation becomes increasingly important.*
>
> —Joseph Bates, A.I. scientist and entrepreneur

When my daughter was little and we were doing something challenging for her—a very long walk or a difficult puzzle, I'd talk to her about stamina. When she first heard the word, she couldn't pronounce it and called it "stanima." It became kind of a catchphrase for us and one I thought of often during my fifteen years with Fairplay. To this day, I hear echoes of "stanima, stanima, stanima" whenever I think about working to prevent corporate marketing to children. Like most worthwhile accomplishments, meaningful social change doesn't come quickly or easily.

Given the centrality of profit to conglomerates targeting kids, it's clear we can't stop, or even significantly curb, marketing to children without government regulation on national, state, and local levels. But policies and practices that promote social justice don't usually emanate from the top. In a society where political candidates rely on corporate donations to fund their campaigns and tech and entertainment companies routinely spend millions each year lobbying elected officials,[1] proposed legislation that threatens corporate profits is unlikely to become law without grassroots efforts that start—often years before—with a few people deciding to work together to right some wrong.

As the oft-repeated quotation attributed to the twentieth-century cultural anthropologist Margaret Mead goes, "Never doubt that a small group of committed citizens can change the world. Indeed, it's the only thing that ever has." Think about those twelve Quakers in England who, in 1797, decided to end slavery in the British Empire. It took them forty-one years, but they did it.[2] Seventy-two years elapsed between the Seneca Falls Women's Rights Convention in 1848 and the passage of the Nineteenth Amendment, which was ratified by Congress in 1920, giving women in the United States the right to vote.[3] The Stonewall uprising took place in 1969 and same-sex marriage wasn't legalized nationally until 2015.[4] Vincent Harding, in his beautiful book *There Is a River*, links the twentieth-century civil rights movement to the centuries-old rebellions aboard slave ships during the terrible Middle Passage.[5]

Today's digitized, commercialized culture is so pervasive and powerful that its influence on children's values, learning, and relationships is a serious threat—not only to the wellbeing of the kids being targeted but also to the wellbeing of democracy and of the planet. But once we recognize that it's unconscionable to leave children unprotected in the marketplace, we can take steps toward change.

If and how you take any of these steps will depend on your time, resources, passions, and inclinations. It is, however, a terrible mistake to afford Big Tech and big business unfettered license to interfere with children's relationships, values, and learning, or any aspect of their development. Children need to be raised by people who love them, in communities that value them, not by entities that profit from their exploitation. Targeting kids with marketing threatens their healthy development. It exacerbates the problems of children and families whose lives are already stressed and it also creates new problems, even for those fortunate enough to be living what could be relatively stress-free lives.

As I wrote in the introduction, new technologies are developing at breakneck speed in the context of little or no regulation. Machines operating with increasing independence and authority already make life-altering decisions such as which prisoners get parole.[6] And robots already serve as caregivers and companions in nursing homes.[7] Have you heard about Delphi, the computer being trained to make moral decisions? Delphi is still in development, but it's available online for people to submit questions (by the way, it does not think that marketing to children is moral—I checked!).

I don't underestimate the powerful forces invested in manipulating children for profit. They include the biggest corporations in the world, which have deep pockets and long-standing connections with policy makers. The good news for people interested in propelling change on a national or international level is that established organizations like Fairplay, Color of Change, Center for Digital Democracy, Instituto Alana in Brazil, and the 5Rights Foundation in the United Kingdom are all working separately and collectively to change the laws, policies, and practices that favor Big Tech and big business over children. Check out these and other organizations listed in the resources to see how you can get involved.

It's still challenging, but it can be somewhat easier, for individuals or small groups of people to make a difference at the local level. Are you satisfied with your school's policies on edtech or commercialism? Does your school district even have such policies? Parents around the country have had success in preventing schools from using tech products that target kids with advertising or invade their privacy. Take Nora Shine, for instance. Nora is a psychologist who sits on her town's school committee. In 2021, she spearheaded a new policy that prevents exploitative edtech products from being used in classrooms in her town's school system. It all began with Prodigy, the freemium app I describe in chapter 10.

"My kids attend public school in Massachusetts, in the Acton-Boxborough Regional School District," she explains. "My nine-year-old daughter was introduced to Prodigy in her math class. When she started pestering me to buy a premium subscription, I asked her to show me the game. I saw that Prodigy isn't really about learning math. It's mainly just a role-playing game, with flying around, battles, and shopping. Solving a few simple math problems to 'win' a battle doesn't make it an effective math game. For me, the worst part was that, like my daughter, kids were manipulated into wanting to buy a premium membership. I wondered what that would feel like for children whose families couldn't afford it.

"The first thing I did was talk to my daughter's teacher. Her response was merely that 'students should learn to ignore advertising.' So then I went to the school principal, who said that 'teachers should have freedom to explore teaching materials to use in their classrooms.' He suggested that I take my concerns to the district level, which I did. That's when I discovered that there were district administrators who shared my concerns about edtech companies exploiting kids. In fact, they were already developing a policy to expand student privacy protections and to improve equitable representation in classroom materials.

"I worked with district leaders, encouraging stronger language around intrusive and manipulative advertising to students. The school committee passed our policy* and now begins the work to ensure that teachers follow it. With all the new tricks used by

---

*See the appendix for an adapted version of the edtech policy used at the Acton-Boxborough Regional School District in Massachusetts.

corporations, it may not be easy. But I hope that other districts do the same thing."

In her efforts to remove Prodigy and other exploitative edtech programs from her district's schools, Nora also reached out to Fairplay for support. That led to the 2021 complaint Fairplay and other advocacy groups submitted to the Federal Trade Commission in 2021.

Nora was concerned about the inherently exploitative nature of using a freemium like Prodigy in classrooms. Some years earlier, Rachael Stickland encountered a different edtech problem in her children's school district—a proposed cloud-based administrative platform called inBloom.

In 2013, Rachael's children were attending school in Jefferson County, Colorado, when she learned her district signed up to pilot inBloom, the Bill and Melinda Gates Foundation's $100 million edtech tool designed to capture and aggregate an astonishing array of student data, including disciplinary records, medical history, and counseling records. The stated purpose of inBloom was to make it easier for teachers to meet students' individual needs, but the data collected would also be made available to other edtech vendors looking to sell their wares.[8]

Rachael says, "I was concerned about the amount and kinds of data they were going to collect and who was going to have access to it. There was no public debate, vote, or information about the initiative made available to parents leading up to or after the district's decision to sign the agreement with inBloom. I reached out to the superintendent, the board of education, PTA leadership, and the teacher's union in the following months. But I couldn't find any local support.

"I did, however, find allies around the country. These were parents like me who wanted to protect their children's privacy in school. We worked together, shared resources, and sought help and guidance from advocacy groups like Class Size Matters and Fairplay. Support from that community helped me get the press involved, which put pressure on school officials. Eventually, inBloom became so controversial that it developed into an issue in our school board election that fall.

"On the evening the new board members were to be sworn in, the outgoing board voted to reject inBloom and the district's superintendent announced her resignation. Weeks later, the Colorado State Board of Education severed its relationship with inBloom."

Fortunately, small groups of parents were protesting inBloom in every one of the other pilot districts around the country. Protests in so many of these districts were so successful in stopping the proposed platform that the Gates Foundation shut down the project in April 2014.[9]

One lesson to be learned from Nora's and Rachael's experiences is the importance of finding allies. It's hard, often ineffective, and sometimes painful to be a lone voice calling for change. Nora was fortunate. Even though her daughter's teacher and school principal dismissed her concerns, she found district administrators who shared her unease and had the power to do something about it. Rachael's quest for allies was harder. Initially, she couldn't find any local support for her concerns—not from parents, teachers, or administrators. Rather, she found allies by reaching out to the parents protesting inBloom in other pilot districts. In addition, both Nora and Rachael reached out to national organizations for support and guidance.

Today, especially when it comes to concerns about the oversized and underregulated role Big Tech plays in children's lives, it's easier than ever to find allies. Check out Fairplay's Screen Time Action Network, a coalition of more than 1,600 organizations and individuals working to reduce the excessive technology harming kids today. You can find colleagues, obtain resources, learn from experts, and discover ways to amplify your own voice. Roll up your sleeves and join the network's working groups for educators, mental health professionals, parent advocates, and more.

You can also organize a screen-free week celebration in your community, town, school, or organization. It's a great way to raise awareness about excessive screen time, to find people who share your concerns, and to help families who want to cut down on time with tech but have not yet taken the first steps.*

If your town or community doesn't have a local advocacy group, you might want to start your own. This is a fertile time for all kinds of activism on these issues. That means there are models to help you, so you don't have to start completely from scratch. Reach out to existing organizations focused on promoting local efforts, like Turning Life On, in Massachusetts. Some organizations have even been successful working with state and local government officials to pass laws and ordinances. That's what KK Myers and Maree Hampton were able to do. They teamed up in 2019 to found LiveMore, ScreenLess, a nonprofit focused on working for and with young people to promote digital wellbeing—balanced and intentional use of technology.

KK, a veteran high school English teacher in Minnesota, noticed a shift in students' behavior and their wellbeing after 2012, when her school issued every student an iPad. She says, "Over the next few years, I saw a dramatic increase in my students' mental health problems—particularly anxiety and depression. In 2019, the Minnesota Student Survey's reporting of mental health data from middle and high school students supported my observations. But it was reading Jean Twenge's book *iGen* that opened my eyes to the links between the explosion of digital devices and the problems I was seeing.

"Maree Hampton and I had been friends for years, and she shares my concerns. Marie is a health educator, with vast experience in health promotion and youth development programs. We believe, and the data overwhelmingly support our belief, that the overuse and misuse of screen time affects children's health,

---

*Visit screenfree.org for information, suggestions, and materials.

wellbeing, and learning. We also believe that everyone, including young people, has a role to play in promoting digital wellness.

"Early on, we formed a youth council and created videos featuring high school students talking about how their experiences with digital media affect their wellbeing. In January 2020, we presented at a Minnesota County forum for elected officials on adolescent mental health. Afterward, a state legislator approached us saying, 'We need a bill. Can you write a bill?' We never had. But we researched how to do it and dove in."

In 2021, with bipartisan authors and co-sponsors, The Digital Wellbeing Bill passed the Minnesota State Legislature, providing $1 million over two years to develop, implement, and evaluate four projects to promote digital wellbeing in Minnesota: an online library of resources for parents, educators, and young people; a statewide communications campaign; education and training for adults; and a peer education and leadership program created for and with young people to support intentional and balanced use of devices.

One thing that struck me during my conversation with KK is that while she and Maree share a common goal, they bring different skills, experience, networks, and expertise to their work benefits their collaboration. In my work building Fairplay, I found that having a clear idea of my own strengths and limitations meant that I knew when someone else's skills or expertise would benefit the organization. And I also found that building relationships outside the organization was crucial—for getting our message out, for deepening my understanding of the issues, and for obtaining help when we needed it.

I was also struck by how KK and Maree responded when that state legislator asked if they'd ever written a bill. They hadn't, but that didn't deter them. They were able to find the help and support they needed to write one. We could say it was lucky they encountered a legislator who listened to their presentation and saw the need for a bill, but what's more important is that they had the commitment, skills, and fortitude to make the most of that encounter.

Finally, it's impressive that, from its early days, LiveMore, ScreenLess has given young people a voice. Each time KK and Maree testified in support the Digital Wellness bill, or met with legislators, they were joined by youth council members who told their own stories about how screen time affects their lives. The authenticity of young people choosing to speak out on social issues that affect them can be powerfully convincing. We can't expect younger children, whose capacities for abstract thinking and impulse control are still in the early stages of development, to be able to look critically at their experience with tech. But it is possible for teenagers to do that. They can be a potent and influential force in educating peers, parents, and younger kids.

It's encouraging that teens are speaking out and raising their own concerns about the influence Big Tech has on their lives.[10] Some young people have formed their own support and advocacy groups, such as the Log Off movement.[11] That young people are working to resist the more harmful business practices of major technology companies bodes well for a future generation of adults who recognize the importance of curbing the influence Big Tech has on the lives of children and want to do something about it.

Of course, most people don't have the time or resources to start an organization or to devote themselves full time to advocacy. But there are lots of ways to get involved at various levels. You can help raise awareness by hosting speakers or viewings and discussions of relevant films or videos at your home, school, community center, or place of worship. If you have funds to make charitable donations, you can support established advocacy groups, especially those that make it a policy not to accept corporate funding. These all work in their own way to prevent Big Tech and big business from exploiting children. Most of these advocacy groups engage in public education, and some host conferences to bring together committed individuals and organizations. Some work with legislators and regulatory agencies and engage in corporate campaigns. Others focus on helping parents manage tech use in their families and support activities known to be beneficial to children.

If you belong to a professional organization, work with like-minded colleagues to try to stop your organization from taking corporate funding or, at the very least, from taking funding from companies that profit from marketing to children. Partnerships between corporations and professional organizations can create unsurmountable barriers to change. How, for instance, can the National Parent Teacher Association take an objective stance on the costs and benefits of educational technology when its "proud national sponsors" include Google and TikTok?[12] Under the heading "Membership Offers," the PTA's website states that it "does not endorse any commercial product or services." Yet, the very next sentence reads, "Companies making a financial contribution to PTA may be entitled to promotional consideration and, in some cases, may have limited use of PTA's marks and assets." Doesn't promotion of a product, let alone use of its logo, signal an endorsement?

Professional organizations in the fields of health and education provide a great service by creating materials, or distributing existing ones, that help parents and prospective parents, as well as educators and day care providers, make thoughtful, developmentally appropriate decisions about how children spend their time. One terrific example is the American Speech-Language-Hearing Association's *Be Tech Wise with Baby!*, which focuses on how to help babies develop communication skills, encourages parents to talk to kids from birth, and includes suggestions for delaying screen time.

If you are involved with a church, synagogue, mosque, or temple, share your concerns with clergy or lay leaders. The values embedded in marketing messages and promoted by social media, such as materialism, greed, and self-indulgence, are antithetical to the values of most mainstream religions. That's why I have long been surprised and dismayed at the paucity of religious and spiritual leaders who are actively involved in efforts to end marketing to children. But that may be changing.

In January 2022, seventy-five leaders representing ten religious and spiritual traditions sent a public letter to Mark Zuckerberg urging him to end Instagram for Kids because the social network "would prove disastrous for children's spiritual development."[13]

The letter describes actions across various faith communities that reflect this sentiment: "Muslims are reducing smartphone use during Ramadan, some Catholic schools are banning screens as modes of education, monasteries are hosting tech-mindfulness meditation sessions, Jewish families are practicing a screen-free Tech Shabbat, and Pope Francis has officially warned the Church of the impacts of screen time. It is undeniable that our spiritual apprehensions represent heartfelt beliefs common broadly to communities of faith and must be duly considered." I was heartened to see these efforts. Like professional organizations, religious and spiritual communities are well-positioned to facilitate grassroots efforts and raise public awareness.

One important lesson I learned during my tenure at Fairplay is the need to articulate a positive vision for the future as well as a critique of the status quo. People would ask me questions such as, "How does a world free from child-targeted marketing differ from what we have now?" Or, more bluntly, "I get what you're against, but what are you *for*?" Here's my answer: I am for a world where children are universally valued for who they are, not for what they or their parents can buy. Where family and community values no longer compete with commercial values for precedence in children's lives. Where kids have lots of "in the real world" time with their friends and with the adults who love and care for them. Where their friendships can flourish without interference from, and monetization by, tech and media companies. Where tech and media programming is designed to enrich children's lives, not corporate coffers. Where their experience of nature is actual and unbranded. Where opportunities for wonder abound. Where what and how kids are taught in school promotes learning, critical thinking, and democratic citizenship, not consumerism and unthinking brand loyalty.[14]

I've been fortunate, in writing this book, to have wide-ranging conversations with all sorts of people about Big Tech and big business in the lives of children. Some shared their immediate concerns and some shared worries about the future. One tech executive told me that he can envision a future fifty years from now, but not beyond that. While machines will never be smarter than humans, they are likely to become, he acknowledged, more powerful than us. Once that happens, he has no idea what the world will be like. And neither do I.

But I do know that the children of today, living as adults in this future world, will be much better off if we provide them with opportunities that nurture and bring to fruition qualities like curiosity, empathy, kindness, creativity, critical thinking, and compassion. If they learn to see through, and resist, the false promises and environmental harms of marketing-induced consumption. If they can recognize the important differences between the real world and what Mark Zuckerberg has now branded the metaverse. None of these qualities or skills can be taught or inspired by corporations looking to exploit kids for profit.

Preventing Big Tech and big business from targeting kids is not a panacea. In and of itself, it won't end hunger, poverty, or war. But here's what it will do. Freeing children from the dictates of commercial culture will afford millions of kids a chance to lead more meaningful lives and to develop the values, abilities, and attributes they need to thrive in the world they will inherit. It's the childhood they deserve.

Afterword

Life goes on. Books end. Events in the overlapping worlds of Big Tech and big business evolve (or devolve) with astonishing rapidity. So, within weeks of my sending off what was supposed to be the absolute final draft of *Who's Raising the Kids*, the European Union voted to regulate tech companies more strictly,[1] even as pundits in the United States expressed doubts that Congress would pass any of the current bills similarly aimed.[2] Meanwhile, Lego and Epic Games, which developed Fortnite, announced their new partnership to lure kids to the metaverse,[3] new evidence emerged detailing Instagram's appalling disregard for the wellbeing of children and teens,[4] and headlines touted the rise of young people pushing back against the tyranny of technology in their lives.[5] In addition, according to a survey by the Pew Research Center, more parents of young children are concerned about the amount of time their kids spend with videogames and on smartphones than they were pre-COVID.[6]

By the time this book is published, the landscape in the long-term struggle between what's best for corporate profits and what's best for children will be changed. But the underlying issues—the harms of immersing kids in a culture dominated by greed merged with increasingly seductive technologies—remain constant. And so does the need to do whatever we can to ensure that corporations are not the dominant force shaping what children learn, what they value, and how they relate to the people around them.

Acknowledgments

I'm deeply indebted to four colleagues who are also good friends, Criscillia Benford, Angela Campbell, Josh Golin, and Tim Kasser. I'm grateful to each of you for your critical eye, your wisdom, your integrity, and your unfailing support through the ups and downs of bringing *Who's Raising the Kids?* to fruition.

Thanks to the following people for reading and commenting on drafts as the book evolved: Rheta Rubenstein, Grigory Tovbis, Lynda Paull, Ellen Bates-Brackett, Shaya Gregory-Poku, Ana Lucia Villela, Karen Motylewski, Shara Drew, Rachael Franz, Tamar Paull, Tara Grove, Faith Boninger, Ari Craine, Nora Shine, Rinny Yourman, and Susan Wadsworth.

So many people were generous with both their expertise and their personal experiences: Jean Rogers, David Monahan, Doug Gentile, Pamela Hurst-Della Pietra, Russell Banks, Joe Bates, Alvin Poussaint, Alice Hanscam, Aliza Kopans, Suzanne Kopans, Frannie Shepard-Bates, Genessa Trietsch, Brooke Lowenstein, Jenn Klepesch, Lauren Paer, Melinda Brown, Aliza Kopans, Suzanne Kopans, Sarah Stannard, Shaunelle Curry, Mindy

Holohan, Frannie Shepherd-Bates, Paula Rees, Alissa Hoyt, Alex Stening, LeAnna Heinrich, Lisa Cline, Sara Stannard, Sharon Maxwell, and Linda Zoe Podbros.

It was a pleasure to work with the whole New Press team, especially my editor, Ellen Adler, who believed in this book even when I had doubts, Emily Albarillo, Jay Gupta, and Derek Warker. Thanks to my agent, Andrew Stuart, for his long-time support and guidance. I was fortunate to have two terrific research assistants, Cecilia Wallace and Priscilla Okama. Alice Peck, Michelle Memran, and Joe Kelly were all so helpful in the early stages of writing. Thanks especially to Alice for helping me see that there actually was a book in all the pages I sent her.

The Fairplay staff was always generous in their willingness to answer questions, offer suggestions, and provide documentation when I needed it. The teachers, parents, and children of the Corner Co-op Preschool welcomed me in yet again to observe kids at play. I first encountered the phrase "Is That Hope?" in Crooked Media's newsletter, *What a Day*. It seemed a particularly apt title for Chapter 11. And thanks and a hug to Sasha for cheering me on when I needed it.

Finally, I am grateful to the Linden Place Pod—Sherry Steiner, and David, Rhoda, and Kara Trietsch for providing much-needed good talk, good food, and great fun during our months of being locked down.

Appendix: Model Edtech Policy for School Districts

Letter to Teachers About a New Edtech Policy and a Digital Tools Criteria Checklist*

Hello AB Educators,

As our district moved to a K–12 one-to-one digital learning environment, the edtech and teaching learning departments collaborated with educators to identify and assess a collection of digital tools that best meet teaching and learning goals and support the needs of each student. Recognizing the hard work all our educators have done this year to engage students and the new realities of digital learning, we want to share with you a ***newly established process for selecting and integrating digital technologies.***

---

*Adapted with permission from Acton-Boxborough Regional School District, Deborah Bookis, EdD, Assistant Superintendent for Teaching and Learning, Amy Bisiewicz, Med, Director of Educational Technology, Peggy Harvey, Med, PK–12 Digital Learning Coordinator.

Moving from the days when educators independently discovered, experimented, and tried out a wide variety of free digital tools, we are now in a place that allows for a cohesive and intentional review and funding of the best available instructional technologies. This evolution aligns with recommendations from federal, state, and nonprofit educational organizations about evaluating tools based on student safety, privacy, and cultural inclusivity. The resulting Digital Toolbox provides a reference for all educators to know which tools can be used for specific learning goals and which tools are not allowed because of their problematic features or marketing practices.

Our digital toolbox promotes the following goals.

- Provide equitable access to digital tools.
- Align tools with state frameworks and district teaching and learning goals.
- Meet student data privacy regulations under the Children's Online Privacy Protection Act (COPPA), which can be confirmed through the Massachusetts Student Privacy Alliance.
- Avoid marketing and advertising that is directed at children and is intrusive or manipulative.
- Avoid "freemium" tools that have limited functionality and/or require payment after a trial period.
- Comply with federal copyright laws.
- Support inclusive pedagogical practice.
- Be culturally responsive and not discriminate on the basis of race, color, sex, sexual orientation, gender identity, religion, disability, ancestry, or national or ethnic origin.
- Develop and provide targeted professional learning.
- Provide timely tech support.
- Consider district funds and capacity to continue supporting paid online subscriptions.

Process for Requesting and Reviewing Tools

Tools not currently included in the Digital Toolbox can be considered via a new Digital Tools Request Form that is posted on our Digital Literacy website. **Please note that even free tools must be approved before classroom implementation.** This checklist can help you know the criteria that must be considered when requesting new digital tools.

Please let us know if you have any questions, and, always, how we can best support the work you do.

Warmly,
The Edtech and Teaching and Learning Departments

Digital Tools Criteria Checklist*

Acton-Boxborough Regional School District

In an effort to ensure that our educators and students have equitable access to digital tools that align with district teaching and learning goals, all tools that are on the Digital Tool Kit have met the following criteria. (Educational digital tools are constantly evolving. We do our best to update this list with available information.) These criteria also apply to any new resources. *Before* requesting digital instructional tools not listed in the Digital Toolbox, either free or requiring paid subscriptions, please use this checklist:

- [ ] Demonstrates evidence of curriculum alignment and can be purposefully integrated
- [ ] Meets student data privacy regulations under Children's Online Privacy Protection Rule (COPPA) that can be confirmed through the Massachusetts Student Privacy Alliance
- [ ] Aims to be free from marketing and advertising that is directed at children, intrusive, or manipulative
- [ ] Does not duplicate readily available online resources currently supported by the district
- [ ] Supports inclusive pedagogical practices
- [ ] Is culturally responsive and does not discriminate on the basis of race, color, sex, sexual orientation, gender identity, religion, disability, ancestry or national or ethnic origin
- [ ] Does not violate copyright laws
- [ ] Does not employ "freemium" pricing strategies or other forms of deceptive marketing that incentivize staff or students to pay for additional features for the tool to be functional
- [ ] Is functional as a free tool or can be funded by the district

---

*Adapted with permission from the Acton-Boxborough Regional School District, Amy Bisiewicz, Deborah Bookis, Peggy Harvey, district administrators.

Suggested Reading, Viewing, and Listening

Reading

Benjamin, Ruha. *Race After Technology.* Cambridge, UK: Polity Press, 2019.

Brickman, Sophie. *Baby Unplugged: One Mother's Search for Reason and Sanity in the Digital Age.* New York: HarperOne, 2021.

Cantor, Patricia A., and Mary M. Cornish. *Teachwise Infant and Toddler Teachers: Making Sense of Screen Media for Children Under Three.* Charlotte, NC: Information Age Publishing, 2016.

Clement, Joe, and Matt Miles. *Screen Schooled: Two Veteran Teachers Expose How Technology Overuse Is Making Our Kids Dumber.* Chicago: Chicago Review Press Incorporated, 2018.

Dunckley, Victoria L. *Reset Your Child's Brain: A Four-Week Plan to End Meltdowns, Raise Grades, and Boost Social Skills by Reversing the Effects of Electronic Screen-Time.* Novato, CA: New World Library, 2015.

Ewen, Stuart. *Captains of Consciousness: Advertising and the Social Roots of the Consumer Culture.* New York: McGraw-Hill, 1977.

Freed, Richard. *Wired Child: Debunking Popular Technology Myths.* North Charleston, SC: CreateSpace Independent Publishing, 2015.

Gonick, Larry, and Tim Kasser. *Hyper-Capitalism: The Modern Economy, Its Values, and How to Change Them.* New York: New Press, 2018.

Hains, Rebecca C., and Nancy A. Jennings, eds. *The Marketing of Children's Toys.* Cham, Switzerland: Palgrave Macmillan, 2021.

Harding, Vincent. *There Is a River: The Black Struggle for Freedom.* New York: Harcourt Brace Jovanovich, 1981.

Hill, Jennifer. *How Consumer Culture Controls Our Kids: Cashing in on Conformity.* Santa Barbara, CA: Praeger, 2016.

Hochschild, Adam. *Bury the Chains: Prophets and Rebels in the Fight to Free an Empire's Slaves.* Boston: Houghton Mifflin, 2005.

Kasser, Tim. *The High Price of Materialism.* Cambridge, MA: MIT Press, 2003.

Linn, Susan. *The Case for Make Believe: Saving Play in a Commercialized World.* New York: New Press, 2008.

Linn, Susan. *Consuming Kids: The Hostile Takeover of Childhood.* New York: New Press, 2004.

Molnar, Alex, and Faith Boninger. *Sold Out: How Marketing in School Threatens Children's Well-Being and Undermines Their Education.* Lanham, MD: Rowman and Littlefield, 2015.

Noble, Safiya Umoja. *Algorithms of Oppression: How Search Engines Reinforce Racism.* New York: New York University Press, 2018.

Norris, Trevor. *Consuming Schools: Commercialism and the End of Politics.* Toronto: University of Toronto Press, 2011.

Plante, Courtney, Craig A. Anderson, Johnie J. Allen, Christopher L. Groves, and Douglas A. Gentile. *Game On! Sensible Answers About Video Games and Media Violence.* Ames, Iowa: Zengen, 2007.

Schor, Juliet B. *Born to Buy: The Commercialized Child and the New Consumer Culture.* New York: Scribner, 2004.

Turkle, Sherry. *Alone Together: Why We Expect More from Technology and Less from Each Other.* New York: Basic Books, 2011.

Turkle, Sherry. *Reclaiming Conversation: The Power of Talk in a Digital Age.* New York: Penguin, 2015.

Wu, Tim. *The Attention Merchants: The Epic Scramble to Get Inside Our Heads.* New York: Knopf, 2016.

Zomorodi, Manoush, *Bored and Brilliant: How Spacing Out Can Unlock Your Most Productive and Creative Self.* New York: St. Martin's Press, 2017.

Zuboff, Shoshana. *The Age of Surveillance Capitalism.* London: Profile, 2019.

Viewing

Kantayya, Shalini, dir. *Coded Bias.* 2020, Brooklyn, NY: 7th Empire Media.

Orlowski, Jeff, dir. *The Social Dilemma.* 2020, Exposure Labs Productions.

Rossini, Elena, dir. *The Illusionists.* 2016, Media Education Foundation (Distributor).

Westbrook, Adam, Lucy King, and Jonah M. Kessel. *The No Good, Very Bad Truth About the Internet and Our Kids.* 2021, New York Times.

Listening

Honold, Lisa, *Unplug and Plug In*, podcasts.apple.com/us/podcast/unplug-and-plug-in/id1561352061.

Lansbury, Janet, *Unruffled*, www.janetlansbury.com/podcast-audio.

Reid, Kisha, *Defending the Early Years Podcast*, dey.org/early-childhood-education-podcasts.

Resources

Each in their own way, the organizations listed below work to curb the power of Big Tech and big business to interfere with children's healthy development. To my knowledge, none of the organizations listed partner with or accept donations from tech or media companies that target children.

Fairplay

fairplayforkids.org
We're a truly independent voice standing up for what kids and families really need. Together, we'll create a world where kids can be kids, free from the false promises of marketers and the manipulations of Big Tech.

Screen Time Action Network at Fairplay

screentimenetwork.org
A global coalition of practitioners, educators, advocates, activists, parents, and caregivers working to promote a healthy childhood by reducing the amount of time kids spend with digital media.

Screen Free Week

screenfree.org
An annual invitation to play, explore, and rediscover the joys of life beyond ad-supported screens. During the first week of May, thousands of families, schools, and communities around the world will put down their entertainment screens for seven days of fun, connection, and discovery.

5Rights Foundation

5rightsfoundation.com
5Rights Foundation exists to make systemic changes to the digital world to ensure it caters for children and young people, by design and default.

Accountable Tech

accountabletech.org
Social media giants are eroding our consensus reality and pushing democracy to the brink. Accountable Tech is fighting back.

Alana Institute

alana.org.br
A socioenvironmental impact organization that promotes children's rights to integral development and fosters new forms of wellbeing.

Algorithmic Justice League

ajl.org
The Algorithmic Justice League is an organization that combines art and research to illuminate the social implications and harms of artificial intelligence.

Backyard Basecamp

backyardbasecamp.org
(Re)connecting Black, Indigenous, and People of Color (BIPOC) to land and nature in Baltimore City.

Center for Digital Democracy

democraticmedia.org
CDD works to protect and expand digital rights and data justice through research-led initiatives designed to influence policy makers, corporate leaders, the news media, civil society, and the general public.

Center for Humane Technology

humanetech.org
We reframe the insidious effects of persuasive technology, expose the runaway systems beneath, and deepen the capacity of global decision makers and everyday leaders to take wise action.

Children and Media Australia

childrenandmedia.org.au
Supporting children's healthy development and protecting their rights and interests as digital and screen media users by assisting families and children's professionals and by influencing decision makers.

Children and Nature Network

childrenandnature.org
We support and mobilize leaders, educators, activists, practitioners, and parents working to turn the trend of an indoor childhood back out to the benefits of nature—and to increase safe and equitable access to the natural world for all.

Children and Screens: Institute for Digital Media and Child Development

childrenandscreens.org
Our vision is to understand and address compelling questions regarding media's impact on child development through interdisciplinary dialogue, public information, and rigorous, objective research bridging the medical, neuroscientific, social science, education, and academic communities.

Color of Change

colorofchange.org
Color of Change is the nation's largest online racial justice organization. As a national online force driven by seven million members, we move decision makers in corporations and government to create a more human and less hostile world for Black people in America.

Defending the Early Years

dey.org
Defending the Early Years (DEY) is a nonprofit organization working for a just, equitable, and quality early childhood education for every young child. DEY advocates for strong economic and social safety nets for all children because equitable education can only occur when society meets children's basic needs for whole health and wellbeing.

Digital Wellness Institute

digitalwellnessinstitute.com
The Digital Wellness Institute exists to equip leaders and change makers with tools to assess and address digital wellness in order to foster a more positive digital culture around the world.

Foolproof Foundation

foolprooffoundation.org
The Foolproof Foundation teaches kids and adults to make healthy skepticism a habit and to question all marketing and advertising. The Foundation's middle and high school curriculums feature peer-to-peer teaching in web-driven modules.

Let Grow

letgrow.org
Let Grow is creating a new path for parents, schools, and America itself — a path back to letting kids have some adventures, make things happen, and develop some self-reliance.

LiveMore, ScreenLess

LiveMoreScreenLess.org
LiveMore, ScreenLess advocates and promotes digital wellbeing for and with young people.

Log Off

logoffmovement.org
Log Off is a movement dedicated to rethinking social media by teens for teens.

Look up

lookup.live
Through our flagship programs and events, we discover, empower, and mobilize youth leaders who are taking action to raise awareness, inspire, and design a healthy, more inclusive, and responsible digital world.

Parents Together

parentstogether.org
Dedicated to organizing and empowering parents.

Raffi Foundation for Child Honoring

raffifoundation.org
Child honoring is a unique social change revolution, one with the child at its heart. It is a positive vision that holds the primacy of early years as key to activating the powerful potential of our species.

Ranking Digital Rights

rankingdigitalrights.org
We evaluate and rank twenty-six of the world's most powerful digital platforms and telecommunications companies on their disclosed policies and practices affecting users' rights to freedom of expression and information and privacy.

Share Save Spend

sharesavespend.org
The mission of Share Save Spend is to help individuals and families develop healthy money habits that honor their values and enhance their financial wellbeing.

Student Data Privacy Project

studentdataprivacyprogject.org
We believe in the importance of keeping kids' data safe and protecting their privacy and their future.

Truce

truceteachers.org

Raising public awareness about the negative effects of violent and stereotyped toys and media on children, families, schools, and society.

Turning Life On

turninglifeon.org

Turning Life On unites and consults with parents, schools, and medical professionals to embrace digital wellness through the implementation of situation-specific, research-based strategies that balance screen time and promote the human experience.

Wait Until 8th

waituntil8th.org

The Wait Until 8th pledge empowers parents to rally together to delay giving children a smartphone until at least eighth grade. By banding together, we will decrease the pressure felt by kids and parents alike over the kids having a smartphone.

Notes

A Note to the Reader

Seth Godin, *All Marketers Are Liars: The Power of Telling Authentic Stories in an Untrusting World* (New York: Portfolio, 2005), 8.

1. Aaron Rupar, "Trump's Friday Night Effort to Weaponize Coronavirus Against His Enemies Has Already Aged Poorly," *Vox*, February 29, 2020.
2. White House Press Conference, "Remarks by President Trump, Vice President Pence, and Members of the Coronavirus Task Force in Press Briefing," The White House, March 24, 2020, trumpwhitehouse.archives.gov/briefings-statements/remarks-president-trump-vice-president-pence-members-coronavirus-task-force-press-briefing-10.
3. An excellent summary of the harms advertising and marketing do to Black and Brown children can be found at "Black Childhood Matters," Campaign for a Commercial-Free Childhood, commercialfreechildhood.org/black-childhood-matters.
4. Safiya Umoja Noble, *Algorithms of Oppression: How Search Engines Reinforce Racism* (New York: New York University Press, 2018); Ruha Benjamin, *The New Jim Code* (Cambridge: Polity Press, 2019).

INTRODUCTION

Children and Screens, "Ask the Experts: Advertising and Kids: Let's Take a (Commercial) Break," YouTube video, November 27, 2020, www.youtube.com/watch?v=lJyDLN4rkbY.

1. Tamar Lewin, "No Einstein in Your Crib? Get a Refund," *New York Times*, October 24, 2009.
2. Sara Miller Llana, "Have the Heirs of Barbie Hit Limit for Risqué Dolls?" *Christian Science Monitor*, May 26, 2006.
3. "NFL Informs Health Advocacy Groups It Will Curb Fantasy Football Marketing to Young Kids," *Fairplay* (blog), July 13, 2016, fairplayforkids.org/nfl-informs-health-advocacy-groups-it-will-curb-fantasy-football-marketing-young-kids.
4. Natasha Singer, Jack Nicas, and Kate Conger, "YouTube Said to Be Fined Up to $200 Million for Children's Privacy Violations," *New York Times*, August 30, 2019.
5. Sheri Madigan et al., "Association Between Screen Time and Children's Performance on a Developmental Screening Test," *JAMA Pediatrics* 173, no. 3 (March 1, 2019): 244–50; Yolanda (Linda) Reid Chassiakos et al., "Children and Adolescents and Digital Media," *Pediatrics* 138, no. 5 (November 1, 2016).
6. "Self-Generated Child Sexual Abuse Material: Attitudes and Experiences in 2020," Thorn, November 14, 2021, info.thorn.org/hubfs/Research/SGCSAM_Attidues&Experiences_YouthMonitoring_FullReport_2021_FINAL%20(1).pdf.
7. Mary Beth Quirk, "Park Service Approves Policy That Allows Corporate Names Inside National Parks," *Consumerist,* January 4, 2017, consumerist.com/2017/01/04/park-service-approves-proposal-to-allow-ads-inside-national-parks.
8. Tim Kasser, *The High Price of Materialism* (Cambridge: MIT Press, 2002).
9. Juliet Schor, "How Consumer Culture Undermines Children's Well-Being," in *Born to Buy: The Commercialized Child and the New Consumer Culture* (New York: Simon and Schuster, 2004), 141; Tim Kasser, "Frugality, Generosity, and Materialism in Children and Adolescents," in *What Do Children Need to Flourish?*, ed. Kristen Anderson Moor and Laura Lippman (New York: Springer, 2005), 371.
10. Alex Hern, "CloudPets Stuffed Toys Leak Details of Half a Million Users," *The Guardian*, February 28, 2017; Lorenzo Franceschi-

Bicchierai, "Hacker Obtained Children's Headshots and Chatlogs from Toymaker Vtech," *Vice*, November 30, 2015.

Chapter 1: What Children Need and Why Corporations Can't Provide It

Kathy Hirsh-Pasek, quoted in *Brain Matters: Putting the First Years First*, directed by Carlota Nelson, brainmattersfilm.com, 2019.

1. Susan Linn, Joan Almon, and Diane Levin, *Facing the Screen Dilemma: Young Children, Technology and Early Education* (Boston: Campaign for a Commercial-Free Childhood; New York: Alliance for Childhood, 2012), PDF.
2. Nicholas Carr, *The Shallows: What the Internet Is Doing to Our Brains* (New York: Norton, 2010), 34.
3. National Academies of Sciences, Engineering, and Medicine, *Communities in Action: Pathways to Health Equity* (Washington, DC: The National Academies Press, 2017).
4. The NPD Group, Retail Tracking Service, U.S.
5. Alexander Kunst, "Available Toys for Children in the Household in the U.S." (Statista, January 6, 2020).
6. See Susan Linn, *The Case for Make Believe: Saving Play in a Commercialized World* (New York: New Press, 2008), 66–67.
7. Linn, *Case for Make Believe*, 68.
8. Jack P. Shonkoff and Deborah A. Phillips, eds., *From Neurons to Neighborhoods: The Science of Early Childhood Development* (Washington, DC: National Academy Press, 2000); Kenneth R. Ginsburg, "The Importance of Play in Promoting Healthy Child Development and Maintaining Strong Parent-Child Bonds," *Pediatrics* 119, no. 1 (January 1, 2007): 182–91, doi.org/10.1542/peds.2006-2697.
9. Walter Loeb, "Geoffrey's Hot Toy List Kicks Off Holiday Sales at Macy's and Toys 'R' Us," *Forbes*, September 28, 2021.
10. Liza Corsillo, "The Top Holiday Toys to Buy Before They Sell Out, According to Toy Experts," *Strategist*, October 1, 2021, nymag.com/strategist/article/top-kids-toys-for-christmas-2021.html.
11. Paul K. Piff et al., "Awe, the Small Self, and Prosocial Behavior," *Journal of Personality and Social Psychology* 108, no. 6 (June 2015): 883–99, doi.org/10.1037/pspi0000018.
12. Walter Isaacson, *Einstein: His Life and Universe* (New York: Simon and Schuster, 2007), 13.

13. Abraham J. Heschel, *Who Is Man?* (Stanford, CA: Stanford University Press, 1965), 81–93.
14. Portions of the last three paragraphs first appeared in *The Case for Make Believe: Saving Play in a Commercialized World*, 193–96.
15. Melanie Rudd, Kathleen D. Vohs, and Jennifer Aaker, "Awe Expands People's Perception of Time, Alters Decision Making, and Enhances Well-Being," *Psychological Science* 23, no. 10 (August, 2012): 1130–36; Paul K. Piff et al., "Awe, the Small Self, and Prosocial Behavior," *Journal of Personality and Social Psychology* 108, no. 6 (2015): 883–99; Nora Davis, "The Role of Transcendent Nature and Awe Experiences on Positive Environmental Engagement" (PhD diss., University of California–Irvine, 2016).
16. Dachar Keltner, *Born to Be Good: The Science of a Meaningful Life* (New York: W.W. Norton, 2009), 268.
17. Rachel Carson, *The Sense of Wonder* (New York: Harper and Row, 1998), 55.
18. Fred Rogers, *Mister Rogers' Neighborhood* (April 7, 1970), episode 1112, www.neighborhoodarchive.com/mrn/episodes/1112/index.html.
19. Fred Rogers, *Mister Rogers' Neighborhood*, episode 1547, aired May 14, 1985, www.neighborhoodarchive.com/mrn/episodes/1547/index.html.
20. World Health Organization–Regional Office for Europe, *Environmental Noise Guidelines for the European Region* (2018), www.euro.who.int/en/health-topics/environment-and-health/noise/publications/2018/environmental-noise-guidelines-for-the-european-region-2018.
21. United Nations Human Rights Council, "Special Rapporteur on Torture and Other Cruel, Inhuman or Degrading Treatment or Punishment" (Report of the Special Rapporteur, Forty-Third Session, Advanced Edited Version, March 20, 2020), 8.
22. Jennifer L. Harris et al., *Fast Food FACTS 2021: Billions in Spending, Continued High Exposure by Youth* (Storrs, CT: University of Connecticut Rudd Center for Food Policy & Obesity, June 2021), PDF.
23. Craig A. Anderson and Brad J. Bushman, "Effects of Violent Video Games on Aggressive Behavior, Aggressive Cognition, Aggressive Affect, Physiological Arousal, and Prosocial Behavior: A Meta-Analytic Review of the Scientific Literature," *Psychological Science* 12, no. 5 (September 2001): 353–59; Anna T. Prescott, James D. Sargent, and Jay G. Hull, "Meta-Analysis of the Relationship Between Violent Video Game Play and Physical Aggression over Time," *Proceedings of the National Academy of Sciences* 115, no. 40 (October 2018): 9882–988.

24. Both *Consuming Kids* and *The Case for Make Believe* have a chapter devoted to sexualization. Also see Diane Levin and Jean Kilbourne, *So Sexy So Soon: The New Sexualized Childhood and What Parents Can Do About It* (New York: Ballantine Books, 2009); Peggy Ornstein, *Cinderella Ate My Daughter: Dispatches from the Front Lines of the New Girlie-Girl Culture* (New York: HarperCollins, 2011); Sharon Lamb and Lynn Mikkel Brown, *Packaging Girlhood: Rescuing Our Daughters from Marketers' Schemes* (New York: St. Martin's, 2007).
25. An example of this exercise can be found on a video of my talk at the first TEDX Pittsburgh, recorded November 14, 2009, on YouTube, www.youtube.com/watch?v=8huWSQKnllE&t=7s.
26. Patricia Marks Greenfield et al., "The Program-Length Commercial," in *Children and Television: Images in a Changing Sociocultural World*, ed. Gordon Berry and Joy Keiko Asamen (Newbury Park, CA: Sage, 1993), 53–72.
27. Jennifer M. Zosh et al., "Talking Shape: Parental Language with Electronic Versus Traditional Shape Sorters," *Mind, Brain, and Education* 9, no. 3 (2015): 136–44.
28. *Encyclopaedia Britannica Online*, s.v. "Hello Kitty," August 28, 2014.
29. Kyung Hee Kim, "The Creativity Crisis: The Decrease in Creative Thinking Scores on the Torrance Tests of Creative Thinking," *Creativity Research Journal* 23, no. 4 (October 1, 2011): 285–95.
30. Dale Kunkel, "Children and Television Advertising," in *Handbook of Children and the Media*, ed. Dorothy G. Singer and Jerome L. Singer (Thousand Oaks, CA: Sage, 2001), 387–88; Irvin Molotsky, "Reagan Vetoes Bill Putting Limits on TV Programming for Children," *New York Times*, November 7, 1988.
31. D.C. Dennison, "The Year of Playing Dangerously," *Boston Globe Magazine*, December 8, 1985, 14–16.

CHAPTER 2: WHO WINS THE GAMES TECH PLAYS?

Adam Alter, *Irresistible: The Rise of Addictive Technology and the Business of Keeping Us Hooked* (New York: Penguin, 2017), 8.

1. "Real Time Billionaires," *Forbes*, www.forbes.com/real-time-billionaires. These are updated every five minutes daily when the stock markets are open. When I first checked in November 2020, Jeff Bezos and Bill Gates ranked number 1 and 2 respectively and Mark Zuckerberg ranked number 4. By March 2022, Mark Zuckerberg dropped down

to the teens and Elon Musk, Jeff Bezos, and Bill Gates were 1, 2, and 4 respectively.
2. See for instance, *American Heritage Medical Dictionary* (Boston: Houghton Mifflin, 2007), medical-dictionary.thefreedictionary.com /technophobe; *Cambridge English Dictionary*, Cambridge University Press, dictionary.cambridge.org/dictionary/nstagr/technophobe; Merriam-Webster.com Dictionary, Merriam-Webster, www.merriam -webster.com/dictionary/technophobia.
3. Rachel Rabkin Peachman, "Mattel Pulls Aristotle Children's Device After Privacy Concerns," *New York Times*, October 5, 2017.
4. Rabkin Peachman.
5. Rabkin Peachman.
6. Rabkin Peachman.
7. Harry Harlow and Robert R. Zimmermann, "Affectional Responses in the Infant Monkey," *Science* 130, no. 3373 (1959): 421–32, www.jstor .org/stable/1758036.
8. Lauren Slater, "Monkey Love: Harry Harlow's Classic Primate Experiments Suggest That to Understand the Human Heart You Must Be Willing to Break It," *Boston Globe*, March 21, 2004.
9. Felix Gillette, "Bringing Home Baby's First Virtual Assistant," *Bloomberg*, January 3, 2017.
10. "Mattel's Nabi Brand Introduces First-Ever Connected Kids Room Platform in Tandem with Microsoft and Qualcomm—Aristotle Mattel, Inc.," corporate.mattel.com/news/mattel-s-nabiR-brand-introduces-first -ever-connected-kids-room-platform-in-tandem-with-microsoft-and -qualcomm-aristotleTM.
11. "Experts and Advocates Ask Mattel to Stop AI 'Aristotle' from Spying on Babies and Kids," *Fairplay* (blog), May 9, 2017, fairplayforkids .org/experts-and-advocates-ask-mattel-stop-ai-aristotle-spying-babies -kids; "Advocates Commend Mattel for Scrapping 'Aristotle' AI Device for Babies and Children," *Fairplay* (blog), October 27, 2017, fairplay forkids.org/advocates-commend-mattel-scrapping-aristotle-ai-device -babies-children; Rabkin Peachman, "Mattel Pulls Aristotle Children's Device After Privacy Concerns."
12. Letter from Senator Ed Markey and Congressman Joe Barton to Margaret H. Georgiadis, CEO, Mattel, September 28, 2017, www.markey .senate.gov/imo/media/doc/Mattel%20letter.pdf.
13. Rabkin Peachman, "Mattel Pulls Aristotle Children's Device After Privacy Concerns."
14. Jonathan B. Wiener and Michael D. Rogers, "Comparing Precaution

in the United States and Europe," *Journal of Risk Research* 5, no. 4 (2002): 317–49.
15. World Commission on the Ethics of Scientific Knowledge and Technology, *The Precautionary Principle* (Paris: UNESCO, 2005), 14.
16. For an excellent discussion on the impact of media and media violence, including violent video games, on children, see Courtney Plante et al., *Game On! Sensible Answers About Video Games and Media Violence* (Ames: Zengen, 2020). Also, Craig A. Anderson et al., "Violent Video Game Effects on Aggression, Empathy, and Prosocial Behavior in Eastern and Western Countries: A Meta-Analytic Review," *Psychological Bulletin* 136, no. 2 (March 2010): 151–73; Craig A. Anderson et al., "Media Violence and Other Aggression Risk Factors in Seven Nations," *Personality and Social Psychology Bulletin* 43, no. 7 (July 2017): 986–98.
17. Anderson et al., *Personality and Social Psychological Bulletin* 136: 151–73.
18. Victoria Rideout and Michael B. Robb, *The Common Sense Census: Media Use by Kids Age Zero to Eight* (San Francisco: Common Sense Media, 2020), 3.
19. Victoria Rideout and Michael B. Robb, *The Common Sense Census: Media Use by Tweens and Teens* (San Francisco: Common Sense Media, 2019).
20. Rideout and Robb, *Media Use by Tweens and Teens*, 3.
21. Rideout and Robb, *Media Use by Tweens and Teens*, 25; Rideout and Robb, *Media Use by Kids Age Zero to Eight*, 18.
22. Matt Richtel, "Children's Screen Time Has Soared in the Pandemic, Alarming Parents and Researchers," *New York Times*, January 16, 2021.
23. *Holding Big Tech Accountable: Legislation to Build a Safer Internet: Hearings Before the Subcommittee on Consumer Protection and Commerce of the Committee on Energy and Commerce*, 117th Congress (2021) (written testimony of Josh Golin, executive director, Fairplay).
24. Rideout and Robb, *Media Use by Kids Age Zero to Eight*, 4.
25. National Association for the Education of Young Children and the Fred Rogers Center, "Technology and Interactive Media as Tools in Early Childhood Programs Serving Children from Birth through Age 8, National Associate for the Education of Young Children," January 2012, www.naeyc.org/sites/default/files/globally-shared/downloads/PDFs/resources/position-statements/ps_technology.pdf.
26. Rideout and Robb, *Media Use by Kids Age Zero to Eight*, 4.
27. Anne Fernald, Virginia A. Marchman, and Adriana Weisleder, "SES Differences in Language Processing Skill and Vocabulary Are Evident

at 18 Months," *Developmental Science* 16 (2013): 234–48, doi:10.1111/desc.12019; K. Ashana Ramsook, Janet A. Welsh, and Karen L. Bierman, "What You Say, and How You Say It: Preschoolers' Growth in Vocabulary and Communication Skills Differentially Predict Kindergarten Academic Achievement and Self-Regulation," *Social Development* 29, no. 3 (2020): 783–800, doi.org/10.1111/sode.12425.

28. Roberta Michnick et al., "(Baby) Talk to Me: The Social Context of Infant-Directed Speech and Its Effects on Early Language Acquisition," *Current Directions in Psychological Science* 24, no. 5 (October 1, 2015): 339–44, journals-sagepub-com.ezp-prod1.hul.harvard.edu/doi/10.1177/0963721415595345.

29. Patricia K. Kuhl, Feng-Ming Tsao, and Huei-Mei Liu, "Foreign-Language Experience in Infancy: Effects of Short-Term Exposure and Social Interaction on Phonetic Learning," *Proceedings of the National Academy of Sciences* 100, no. 15 (July 22, 2003): 9096–9101, doi.org/10.1073/pnas.1532872100.

30. "Be Tech Wise with Baby! Create a Healthy Technology Environment for Your Baby to Thrive," American Speech-Language-Hearing Association and Campaign for a Commercial-Free Childhood, 2020, screentimenetwork.org/sites/default/files/resources/Be%20Tech%20Wise%20With%20Baby%20final%20Engish.pdf.

31. Jenny Radesky et al., "Maternal Mobile Device Use During a Structured Parent–Child Interaction Task," *Academic Pediatrics* 15, no. 2 (March 1, 2015): 238–44.

32. Sherry Madigan et al., "Associations Between Screen Use and Child Language Skills: A Systematic Review and Meta-Analysis," *JAMA Pediatrics* 174, no. 7 (March 23, 2020): 665–75, doi:10.1001/jamapediatrics.2020.0327.

33. Cortney A. Evans, Amy B. Jordan, and Jennifer Horner, "Only Two Hours? A Qualitative Study of the Challenges Parents Perceive in Restricting Child Television Time," *Journal of Family Issues* 32, no. 9 (March 2011): 1223–44.

34. Council on Communications and Media, "Media and Young Minds," *Pediatrics* 138, no. 5 (November 1, 2016): 2016–2591 doi.org/10.1542/peds; see also Bernard G. Grela, Marina Krcmar, and Yi-Jiun Lin, "Can Television Help Toddlers Acquire New Words?," Speechpathology.com, May 17, 2004, www.speechpathology.com/Articles/article_detail.asp?article_id=72; see also Patricia K. Kuhl, Feng-Ming Tsao, and Huel-Mel Liu, "Foreign-Language Experience in Infancy: Effects of Short-Term Exposure and Social Interaction," *Proceedings of the*

National Academy of Science 100 (2003): 9096–9101; Daniel R. Anderson and Tiffany A. Pempek, "Television and Very Young Children," *American Behavioral Scientist* 48, no. 5 (2005): 505–22.

35. Council on Communications and Media, *Pediatrics* 138: 2016-2591; See also, Ana Maria Portugal et al., "Saliency-Driven Visual Search Performance in Toddlers with Low- vs High-Touch Screen Use," *JAMA Pediatrics* 175, no. 1 (January 1, 2021): 96–97; Anderson and Pempek, *American Behavioral Scientist* 48: 505–22.
36. Elizabeth A. Vandewater, David S. Bickham, and June H. Lee, "Time Well Spent? Relating Television Use to Children's Free-Time Activities," *Pediatrics* 117, no. 2 (2006): 181–91.
37. Vandewater, Bickham, and Lee, *Pediatrics* 117: 181–91.
38. Ana Maria Portugal et al., "Longitudinal Touchscreen Use Across Early Development Is Associated with Faster Exogenous and Reduced Endogenous Attention Control," *Scientific Reports* 11, no. 1 (2021): 2205. Doi.org/10.1038/s41598-021-81775-7.
39. Dylan P. Cliff et al., "Early Childhood Media Exposure and Self-Regulation: Bidirectional Longitudinal Associations," *Academic Pediatrics* 18, no. 7 (September 1, 2018): 813–19, doi.org/10.1016/j.acap.2018.04.012.
40. Laura K. Certain and Robert S. Kahn, "Prevalence, Correlates, and Trajectory of Television Viewing Among Infants and Toddlers," *Pediatrics* 109 (2002): 634–42; Aletha C. Huston et al., "Development of Television Viewing Patterns in Early Childhood: A Longitudinal Investigation," *Developmental Psychology* (1990): 409–20; Dimitri A. Christakis and Frederick Zimmerman, "Early Television Viewing Is Associated with Protesting Turning Off the Television at Age 6," *Medscape General Medicine* 8, no. 2 (2006): 63, www.medscape.com/viewarticle/531503.
41. Jenny S. Radesky and Dimitri A. Christakis, "Increased Screen Time: Implications for Early Childhood Development and Behavior," *Pediatrics Clinics of North America* 65, no. 5 (October 2016): 827–39, doi.org/10.1016/j.pcl.2016.06.006.
42. Marjorie J. Hogan and Victor C. Strasburger, "Media and Prosocial Behavior in Children and Adolescents," in *Handbook of Moral and Character Education*, ed. Larry Nucci and Daria Narvaez (Mahwah, NJ: Lawrence Erlbaum, 2008), 537–53.
43. Patricia Greenfield, "Technology and Informal Education: What Is Taught, What Is Learned," *Science* 323, no. 5910 (2009): 69–71.
44. Council on Communications and Media, *Pediatrics* 138 (2009): 2016-2591.

45. Council on Communications and Media, *Pediatrics* 138: 2016-2591; Council on Communications and Media, "Media Use in School-Aged Children and Adolescents," *Pediatrics* 138, no. 5 (November 1, 2016), doi.org/10.1542/peds.2016-2592.
46. Aric Sigman, "Virtually Addicted: Why General Practice Must Now Confront Screen Dependency," *British Journal of General Practice* 64, no. 629 (December 1, 2014): 610-11, doi.org/10.3399/bjgp14X682597.
47. "Advocates Ask American Psychological Association to Condemn Tactics Used to Hook Kids on Screen Devices," Campaign for a Commercial-Free Childhood, August 6, 2018, commercialfreechildhood.org/advocates-ask-american-psychological-association-condemn-tactics-used-hook-kids-screen-devices.
48. Susan Weinschenk, "The Dopamine Seeking-Reward Loop," *Psychology Today*, February 28, 2016, www.psychologytoday.com/blog/brain-wise/201802/the-dopamine-seeking-reward-loop; Simon Parkin, "Has Dopamine Got Us Hooked on Tech?," *The Guardian*, March 4, 2018.
49. Transparency Market Research, "Smart Toys Market to Observe Stellar Growth in Future with Growth Projected at Whopping 36.4% CAGR for 2018-2026, Integration of Futuristic Internet of Toys Technology to Open Large Growth Vistas-TMR," *PR Newswire*, April 13, 2021.
50. Linda Neville, "Kids' Brand Must Exercise Pest Control," *Brand Strategy*, November 2, 2001, 17.
51. Craig Timberg, "The YouTube Conundrum: Site Says It Bans Preteens, but Children Are Still Able to Stream Troubling Content with Very Little Effort," *Owen Sound Sun Times*, March 29, 2019, advance-lexis-com.ezp-prod1.hul.harvard.edu/api/document?collection=news&id=urn:contentItem:5VS2-55V1-JC4X-J541-00000-00&context=1516831.
52. K.G. Orphanides, "Children's YouTube Is Still Churning Out Blood, Suicide and Cannibalism," *Wired UK*, March 23, 2018, www.wired.co.uk/article/youtube-for-kids-videos-problems-algorithm-recommend.
53. Madhav, "Fortnite Business Model: How Does Fortnite Make Money?," *SEOAves*, June 13, 2021, seoaves.com/how-does-fortnite-make-money-fortnite-business-model.
54. Brett Hershman, "7 Crazy 'Fortnite' Stats—and How Virtual Fashion Is Driving In-Game Purchases," *Benzinga*, September 6, 2018, www.benzinga.com/news/18/09/12317014/7-crazy-fortnite-stats-and-how-virtual-fashion-is-driving-in-game-purchases.

55. "How Much Money Does Fortnite Make? 2020 Revenue Revealed," *Elecspo* (blog), May 4, 2021, www.elecspo.com/games/fortnite/how-much-does-fortnite-make.
56. Richard Freed, "The Tech Industry's War on Kids," *Medium*, March 12, 2018, medium.com/@richardnfreed/the-tech-industrys-psychological-war-on-kids-c452870464ce.
57. For a good history of advertising, the role of psychology, and the elevation of consumer culture, see Stuart Ewen, *Captains of Consciousness: Advertising and the Social Roots of the Consumer Culture* (New York: Basic Books, 2008).
58. See Dan S. Acuff and Robert H. Reiher, *What Kids Buy and Why: The Psychology of Marketing to Kids* (New York: The Free Press, 1997), 16, for a description of the links between developmental psychology and successful marketing to children.
59. Brian L. Wilcox et al., "Report of the APA Task Force on Advertising and Children," American Psychological Association, February 20, 2004, www.apa.org/pubs/info/reports/advertising-children.
60. "Advertising Recommendations," *American Psychological Association* 35, no. 6 (June 4, 2004): 59, www.apa.org/monitor/jun04/apatask.
61. "Our Letter to the American Psychological Association," Children's Screen Time Action Network, August 8, 2018, screentimenetwork.org/apa.
62. Freed, *Medium*, March 12, 2018.
63. Genxee, "7 Years Later: Minecraft $2.5B Deal Microsoft's Most Successful Acquisition?," *Influencive*, June 30, 2021, www.influencive.com/7-years-later-minecraft-2-5b-deal-microsofts-most-successful-acquisition.
64. Globe Newswire, "ThinkGeek and Mojang Build on Licensing Deal," news release, *Dow Jones Institutional News*, June 15, 2011.
65. Alex Cox, "The History of Minecraft—the Best Selling PC Game Ever," *TechRadar*, September 4, 2020, www.techradar.com/news/the-history-of-minecraft.
66. Sandra Fleming, "The Parents Guide to the Minecraft Marketplace," *Best Apps for Kids* (blog).
67. Sherry Turkle, "Opinion: There Will Never Be an Age of Artificial Intimacy," *New York Times*, August 11, 2018.

Chapter 3: And the Brand Plays On

Robert Iger, quoted in Gary Gentile, "Toy Story: Disney Aims to Capture Elusive Boys Market," *Pittsburgh Post-Gazette*, November 9, 2004.

1. "PlayCon—Overview," The Toy Association, www.toyassociation.org/toys/events/playcon-home.aspx.
2. Sarah Whitten, "If Toys R Us Liquidates, 10 to 15 Percent of All Toy Sales Could Be Lost Forever," CNBC, March 9, 2018, www.cnbc.com/2018/03/09/10-percent-15-percent-of-all-toy-sales-could-be-lost-forever-if-toys-r-us-liquidates.html.
3. "Tears 'R' Us: The World's Biggest Toy Store Didn't Have to Die," *Bloomberg*, June 6, 2018.
4. Charisse Jones, "Who Are the Winners and Losers After Toys 'R' Us Fall? Walmart, Amazon and Mattel So Far," *USA Today*, July 26, 2018.
5. Kristen Korosec, "Toys 'R' Us Closings: How Mattel, Hasbro, and Lego Will Be Affected," *Fortune*, March 16, 2018.
6. Bryce Covert, "The Demise of Toys 'R' Us Is a Warning," *Atlantic*, July/August 2018.
7. Chavie Lieber, "Thousands of Toys 'R' Us Workers Are Getting Severance, Following Months of Protests," *Vox*, November 21, 2018.
8. Joseph Pereira, "Toys 'R' Us, Big Kid on the Block, Won't Stop Growing: Retailer's Grand-Scale Strategy Works, to the Dismay of Competitors," *Wall Street Journal*, August 11, 1988.
9. Emily Walsh, "Macy's Wants to Hire 76,000 Workers Before the Holidays, Nearing Pre-Pandemic Levels of Employment," *Business Insider, US Edition*, September 22, 2021.
10. Kathleen Elkins, "How James Patterson, the Richest Writer in the World, Helped Create the Iconic Toys 'R' Us Jingle," CNBC, March 19, 2018.
11. "Tears 'R' Us: The World's Biggest Toy Store Didn't Have to Die," *Bloomberg*, June 6, 2018.
12. Derek Thompson, "Who Bankrupted Toys 'R' Us? Blame Private Equity and Millennial Parents," *The Atlantic*, January 24, 2018.
13. Julia Horowiz, "How Toys 'R' Us Went from Big Kid on the Block to Bust," *CNN Business*, March 17, 2018.
14. Michael S. Rosenwald, "Toys 'R' Us: The Birth—and Bust—of a Retail Empire," *Chicago Tribune*, September 19, 2017.
15. "Toys 'R' Us, Inc.," Company-Histories, www.company-histories.com/Toys-R-Us-Inc-Company-History.html.

16. Caroline E. Mayer, "Happy Campers at the Store, Retailers Find Summer Kids Programs Pay Off," *Washington Post*, July 12, 2003.
17. James B. Twitchell, *Lead Us into Temptation* (New York: Columbia University Press, 1999), 30.
18. Ron Harris, "Children Who Dress for Excess: Today's Youngsters Have Become Fixated with Fashion, the Right Look Isn't Enough—It Also Has to Be Expensive," *Los Angeles Times*, November 12, 1989.
19. Richard Fry and Kim Parker, "'Post-Millennial' Generation on Track to Be Most Diverse, Best-Educated," *Pew Research Center's Social and Demographic Trends Project* (blog), November 15, 2018, www.pewresearch.org/social-trends/2018/11/15/early-benchmarks-show-post-millennials-on-track-to-be-most-diverse-best-educated-generation-yet.
20. *Merriam-Webster Word Central*, s.v. "tribe (*n.*)," www.wordcentral.com/cgi-bin/student?book=Student&va=tribe.
21. Amy Chua, *Political Tribes: Group Instinct and the Fate of Nations* (New York: Penguin, 2018).
22. Author's notes, Playcon conference, San Francisco, May 8, 2018.
23. Ranyechi Udemezue, "The Seven Teen Tribes You Need to Know Now," *Sundaytimes.Co.Uk*, February 6, 2021.
24. "Brand Communities and Consumer Tribes," Vivid Brand, vividbrand.com.
25. Tina Sharkey, "What's Your Tribe? Tap into Your Core Consumers' Aspirations Like Nike, Gatorade, BabyCenter and REI Do," *Forbes*, January 25, 2012.
26. "CCFC to Unilever: Ax the Axe Campaign If You Care About 'Real Beauty,'" Campaign for a Commercial-Free Childhood.
27. Urie Bronfenbrenner, "Toward an Experimental Ecology of Human Development," *American Psychologist* 32, no. 7 (July 1977): 513–31, doi.org/10.1037/0003-066X.32.7.513.
28. Katherine Sell et al., "The Effect of Recession on Child Well-Being: A Synthesis of the Evidence by Policy Lab and Children's Hospital of Philadelphia," *Foundation for Child Development*, November 6, 2010.
29. Laura Ly, "Judge Gives Final Approval of $626 Million Settlement for People Affected by Flint Water Crisis," CNN, November 10, 2021, www.cnn.com/2021/11/10/us/flint-michigan-water-crisis-judge-approves-settlement/index.html.
30. "Reputation vs. Brand: What's the Difference?," *Mail and Guardian*, September 30, 2016.

31. Bloomberg Business News, "Brand Loyalty Is the New Holy Grail for Advertisers; Marketing: Making Sure the Customer Keeps Coming Back Is Viewed as Path to Maximum Profit," *Los Angeles Times*, July 18, 1996.
32. Paul M. Connell, Merrie Brucks, and Jesper H. Nielsen, "How Childhood Advertising Exposure Can Create Biased Product Evaluations That Persist into Adulthood," *Journal of Consumer Research* 41, no. 1 (June 1, 2014): 119–34, doi.org/10.1086/675218.
33. Julia Fein Azoulay, "Brand Aware," *Children's Business* 15, no. 6 (June, 200): 46–48.
34. Thomas N. Robinson et al., "Effects of Fast Food Branding on Young Children's Taste Preferences," *Archives of Pediatrics and Adolescent Medicine* 161, no. 8 (August 2007): 792–97, doi.org/10.1001/archpedi.161.8.792.
35. Christina A. Roberto et al., "Influence of Licensed Characters on Children's Taste and Snack Preferences," *Pediatrics* 126, no. 1 (July 1, 2010): 88–93, doi.org/10.1542/peds.2009-3433.
36. Carl F. Mela, Sunil Gupta, and Donald R. Lehmann, "The Long-Term Impact of Promotion and Advertising on Consumer Brand Choice," *Journal of Marketing Research* 34 (1997): 248.
37. Mela, Gupta, and Lehmann, *Journal of Marketing Research* 34: 248.
38. Ben Kamisar, "Trump: I Could Shoot People in Streets and Not Lose Support," *The Hill*, January 23, 2016, thehill.com/blogs/ballot-box/gop-primaries/266809-trump-i-could-shoot-people-in-streets-and-not-lose-support.
39. Neal Larson, "Apple Would Be in Trouble with Customers If They Weren't a Cult," *Idaho State Journal*, December 31, 2017, idahostatejournal.com/opinion/columns/apple-would-be-in-trouble-with-customers-if-they-weren/article_feb63db0-7c47-51a2-a17d-6e95a8ebbf69.html; Lindsay Willott, "10 Brand Loyalty Statistics for 2017," Customer Thermometer, May 25, 2017, www.customerthermometer.com/customer-retention-ideas/brand-loyalty-statistics-2017; Kurt Badenhausen, "The World's Most Valuable Brands 2019: Apple on Top at $206 Billion," *Forbes*, May 22, 2019; Marty Swant, "Apple, Microsoft and Other Tech Giants Top Forbes' 2020 Most Valuable Brands List," *Forbes*, May 22, 2019; Communications, "*Forbes* Releases Seventh Annual World's Most Valuable Brands List," *Forbes*, May 23, 2017.
40. "Apple 1984 Super Bowl Commercial Introducing Macintosh Computer," www.youtube.com/watch?v=2zfqw8nhUwA.

41. "Apple CEO Tim Cook: 'Privacy Is a Fundamental Human Right,'" NPR, October 1, 2015, www.npr.org/sections/alltechconsidered/2015/10/01/445026470/apple-ceo-tim-cook-privacy-is-a-fundamental-human-right.
42. Chen Guangcheng, "Apple Can't Resist Playing by China's Rules," *New York Times*, January 23, 2018.
43. Jack Nicas, Raymond Zhong, and Daisuke Wakabayashi, "Censorship, Surveillance and Profits: A Hard Bargain for Apple in China," *New York Times*, May 17, 2021.
44. Elisabeth Bumiller, "Reagan's Ad Aces," *Washington Post*, October 18, 1984.
45. Matthew Creamer, "Obama Wins! Ad Age's Marketer of the Year," *AdAge*, October 17, 2008, adage.com/article/moy-2008/Obama-wins-ad-age-s-marketer-year/131810.
46. Ellen McGirt, "The Brand Called Obama," *Fast Company*, April 1, 2008.
47. Staci M. Zavattaro, "Brand Obama: The Implications of a Branded President," *Administrative Theory and Praxis* 32, no. 1 (March 2010): 123–28, www.jstor.org/stable/25611043.
48. Robert Schlesinger, "Ka-Ching: Donald Trump Is Raking in Big Bucks from Emoluments Foreign and Domestic," *US News and World Report*, March 5, 2018.
49. Jordan Libowitz, "Profiting from the Presidency: A Year's Worth of President Trump's Conflicts of Interest," CREW, January 19, 2018, www.citizensforethics.org/press-release/crew-releases-report-profiting-presidency-years-worth-president-trumps-conflicts-interest.
50. Reid J. Epstein, "The G.O.P.'s Official Stance in 2020 Is That It Is for Whatever Trump Says," *New York Times*, August 25, 2020.
51. Steve Chapman, "Donald Trump Is a Profoundly Incompetent President," *Chicago Tribune*, June 7, 2017.
52. Kenneth P. Vogel, "Trump Leaves His Mark on a Presidential Keepsake," *New York Times*, June 24, 2018; Tara McKelvey, "How Trump Uses Reagan's Playbook on the White House Lawn," *BBC News*, February 23, 2018, www.bbc.com/news/world-us-canada-42969951; Savannah Rychcik, "Chris Christie Claims Trump Loved 'Trappings' of the Office 'Most Days' More than the Job Itself," *IJR* (blog), January 20, 2021, ijr.com/chris-christie-trump-loved-trappings-of-office/; Ken Thomas and Julie Pace, "Trump Getting Better at Using the Trappings of Office," AP News, March 2, 2017, apnews.com/article/96f175ae266c43bcaadea39c831e13a1.
53. Steve Fogg, "10 Common Branding Mistakes That Churches Make,"

Steve Fogg, November 21, 2012, www.stevefogg.com/2012/11/21/branding-churches.

54. Natasha Singer, "Silicon Valley Courts Brand-Name Teachers, Raising Ethics Issues," *New York Times*, September 2, 2017.
55. Bob Roehr, "Pharma Gifts Associated with Higher Number and Cost of Prescriptions Written," *BMJ* 359 (October 26, 2017), j4979, doi.org/10.1136/bmj.j4979.
56. Paris Martineau, "The WIRED Guide to Influencers," *Wired*, December 6, 2020, www.wired.com/story/what-is-an-influencer.
57. Natalya Saldanha, "In 2018, an 8-Year-Old Made $22 Million on YouTube. No Wonder Kids Want to Be Influencers," *Fast Company*, November 19, 2019, www.fastcompany.com/90432765/why-do-kids-want-to-be-influencers.
58. Madeline Berg and Abram Brown, "The Highest-Paid YouTube Stars of 2020," *Forbes*, December 18, 2020, www.forbes.com/sites/maddieberg/2020/12/18/the-highest-paid-youtube-stars-of-2020/?sh=3185e9a56e50.
59. "Ryan's World—Shop by Category," Target.com, www.target.com/c/ryan-s-world/-/N-nxa8t.
60. Jay Caspian Kang, "Ryan Kaji, the Boy King of YouTube," *New York Times Magazine*, January 9, 2022, 22.
61. Mariska Kleemans et al., "Picture Perfect: The Direct Effect of Manipulated Instagram Photos on Body Image in Adolescent Girls," *Media Psychology* 21, no. 1 (January 2, 2018): 93–110, doi.org/10.1080/15213269.2016.1257392.
62. Jean M. Twenge, Gabrielle N. Martin, and W. Keith Campbell, "Decreases in Psychological Well-Being Among American Adolescents After 2012 and Links to Screen Time During the Rise of Smartphone Technology," *Emotion* 18, no. 6 (2018): 765–80, doi.org/10.1037/emo0000403.
63. Mary Aiken, "The Kids Who Lie About Their Age to Join Facebook," *The Atlantic*, August 30, 2016; "Under-Age Social Media Use 'on the Rise,' Says Ofcom," *BBC News*, November 29, 2017, www.bbc.com/news/technology-42153694; Alise Mesmer, "YouTube Says It Removed over 7M Accounts Belonging to Users Under 13 This Year Alone," *Newsweek*, October 26, 2021.
64. Marika Tiggemann and Amy Slater, "NetTweens: The Internet and Body Image Concerns in Preteenage Girls," *Journal of Early Adolescence* 34, no. 5 (June 1, 2014): 606–20, doi.org/10.1177/0272431613501083.
65. Ryan Mac and Craig Silverman, "Facebook Is Building an Instagram

for Kids Under the Age of 13," *BuzzFeed News*, March 18, 2021, www.buzzfeednews.com/article/ryanmac/facebook-instagram-for-children-under-13.

66. Julie Jargon, "Life and Arts—Family and Tech: Facebook Messenger Kids: A Chat App for Tweens?," *Wall Street Journal*, May 13, 2020.
67. Russell Brandom, "Facebook Design Flaw Let Thousands of Kids Join Chats with Unauthorized Users," *Verge*, July 22, 2019, www.theverge.com/2019/7/22/20706250/facebook-messenger-kids-bug-chat-app-unauthorized-adults.
68. Samantha Murphy Kelly, "Facebook Says It's Moving Forward with Instagram for Kids Despite Backlash," *CNN Business*, July 27, 2021.
69. "House Hearing on Combating Online Misinformation and Disinformation," C-SPAN.Org, March 25, 2021, www.c-span.org/video/?510053-1/house-hearing-combating-online-misinformation-disinformation.
70. Deepa Seetharaman, Georgia Wells, and Jeff Horwitz, "The Facebook Files: Facebook Knows Instagram Is Toxic for Teen Girls, Its Research Shows—Internal Documents Show a Youth Mental-Health Issue that Facebook Plays Down in Public," *Wall Street Journal*, September 15, 2021.
71. Seetharaman, Wells, and Horwitz, "The Facebook Files."
72. Sam Machkovech, "Report: Facebook Helped Advertisers Target Teens Who Feel 'Worthless' [Updated]," Ars Technica, May 1, 2017, arstechnica.com/information-technology/2017/05/facebook-helped-advertisers-target-teens-who-feel-worthless.
73. Dylan Williams, Alexandra McIntosh, and Rys Farthing, "Profiling Children for Advertising: Facebook's Monetisation of Young People's Personal Data," Reset Australia, April 26, 2021. au.reset.tech/news/profiling-children-for-advertising-facebooks-monetisation-of-young-peoples-personal-data.
74. Josh Taylor, "Facebook Allows Advertisers to Target Children Interested in Smoking, Alcohol and Weight Loss," *The Guardian*, April 28, 2021.
75. Williams, McIntosh, and Farthing, "Profiling Children for Advertising."
76. Natasha Singer, "Mark Zuckerberg Is Urged to Scrap Plans for an Instagram for Children," *New York Times*, April 15, 2021.
77. Adam Barnes, "44 US Attorneys General Urge Facebook to Cancel 'Instagram for Kids,'" *The Hill*, May 11, 2021, thehill.com/changing-america/well-being/552797-44-us-attorneys-general-urge-facebook-to-cancel-instagram-for.

78. Senator Ed Markey and Representatives Kathy Castor and Lori Trahan wrote a public letter to Mark Zuckerberg urging him not to go forward with Instagram for Kids. www.markey.senate.gov/imo/media/doc/fb_wsj_report.pdf, September 15, 2021.
79. Adam Mosseri, "Pausing 'Instagram Kids' and Building Parental Supervision Tools," Instagram, September 27, 2021, about.instagram.com/blog/announcements/pausing-instagram-kids.
80. Mosseri, "Pausing 'Instagram Kids.'"
81. Georgia Wells and Jeff Horwitz, "Facebook's Effort to Attract Preteens Goes Beyond Instagram Kids, Documents Show," *Wall Street Journal*, September 28, 2021, www.wsj.com/articles/facebook-instagram-kids-tweens-attract-11632849667.
82. Kevin Roose, "Facebook Is Weaker Than We Knew," *New York Times*, October 4, 2021.

Chapter 4: Browse! Click! Buy! Repeat!

"Amazon CEO Jeff Bezos Tip #37: 'Reduce Friction,'" YouTube, March 10, 2013, www.youtube.com/watch?v=hUtQv8YWCGE.

1. Tim Kasser, *The High Price of Materialism* (Cambridge: MIT Press, 2002).
2. My favorite discussion of this phenomenon is by Donald W. Winnicott, *Playing and Reality* (New York: Basic Books, 1971).
3. Tim Kasser et al., "Some Costs of American Corporate Capitalism: A Psychological Exploration of Value and Goal Conflicts," *Psychological Inquiry* 18, no. 1 (March 1, 2007): 1–22.
4. Committee on Integrating the Science of Early Childhood Development, *From Neurons to Neighborhoods: The Science of Early Childhood Development*, National Academies Press, November 13, 2000.
5. *Merriam-Webster*, s.v. "materialistic (*adj.*)," www.merriam-webster.com/dictionary/materialistic.
6. *Urban Dictionary*, s.v. "materialistic," www.urbandictionary.com/define.php?term=materialistic.
7. For a condensed, research-based overview of materialism, children, and advertising, see Campaign for a Commercial-Free Childhood, "Get the Facts: Marketing and Materialism," www.commercialfreechildhood.org/sites/default/files/devel-generate/wri/materialism_fact_sheet.pdf; for research studies, see Suzanna J. Opree et al., "Children's Advertising Exposure, Advertised Product Desire, and Materialism:

A Longitudinal Study," *Communication Research* 41, no. 5 (July 2014): 717–35, doi.org/10.1177/0093650213479129; Moniek Buijzen and Patti M. Valkenburg, "The Effects of Television Advertising on Materialism, Parent-Child Conflict, and Unhappiness: A Review of Research," *Journal of Applied Developmental Psychology* 24, no. 4 (September 1, 2003): 437–56, doi.org/10.1016/S0193-3973(03)00072-8; Marvin E. Goldberg and Gerald J. Gorn, "Some Unintended Consequences of TV Advertising to Children," *Journal of Consumer Research* 5, no. 1 (June 1, 1978): 22–29, doi.org/10.1086/208710; Vanessa Vega and Donald F. Roberts, "Linkages Between Materialism and Young People's Television and Advertising Exposure in a US Sample," *Journal of Children and Media* 5, no. 2 (April 15, 2011): 181–93, doi.org/10.1080/17482798.2011.558272.

8. For a meta-analysis on the links between materialism and environmental degradation see Megan Hurst et al., "The Relationship Between Materialistic Values and Environmental Attitudes and Behaviors: A Meta-Analysis," *Journal of Environmental Psychology* 36 (2013): 257–69.

9. Judith Stephenson et al., "Population, Development, and Climate Change: Links and Effects on Human Health," *Lancet* 382, no. 9905 (November 16, 2013): 1665–73, doi.org/10.1016/S0140-6736(13)61460-9.

10. Joe Pinsker, "A Cultural History of the Baseball Card," *The Atlantic*, December 17, 2014.

11. See "Happy Meal" in *Wikipedia*, en.wikipedia.org/w/index.php?title=Happy_Meal&oldid=1043595386.

12. Craig Donofrio and Brittany Alexandra Sulc, "Most Valuable Beanie Babies," Work + Money, January 4, 2022, www.workandmoney.com/s/most-valuable-beanie-babies-e902756fef944af3.

13. *Bulbapedia, the Community-Driven Pokémon Encyclopedia*, s.v. "History of Pokémon," bulbapedia.bulbagarden.net/wiki/History_of_Pok%C3%A9mon.

14. Ellen Macarthur Foundation, *The New Plastics Economy: Rethinking the Future of Plastics*, World Economic Forum, January 2016, www3.weforum.org/docs/WEF_The_New_Plastics_Economy.pdf.

15. GrrlScientist, "Five Ways That Plastics Harm the Environment (And One Way They May Help)," April 23, 2018, www.forbes.com/sites/grrlscientist/2018/04/23/five-ways-that-plastics-harm-the-environment-and-one-way-they-may-help/#4b9d6e1567a0; Chris Wilcox, Erik Van Sebille, and Britta Denise Hardesty, "Threat of

Plastic Pollution to Seabirds Is Global, Pervasive, and Increasing," *Proceedings of the National Academy of Sciences of the United States of America* 112, no. 38 (September 22, 2015): 11899–904, doi.org/10.1073/pnas.1502108112; Qamar A. Schuyler et al., "Risk Analysis Reveals Global Hotspots for Marine Debris Ingestion by Sea Turtles," *Global Change Biology* 22, no. 2 (February 2016): 567–76, doi.org/10.1111/gcb.13078; Bianca Unger et al., "Large Amounts of Marine Debris Found in Sperm Whales Stranded Along the North Sea Coast in Early 2016," *Marine Pollution Bulletin* 112, no. 1 (November 15, 2016): 134–41, doi.org/10.1016/j.marpolbul.2016.08.027.

16. Alana Semuels, "The Strange Phenomenon of L.O.L. Surprise Dolls," *The Atlantic*, November 29, 2018.
17. "Toymaker Opens Up About the Season's Hot New Toy," CBS News, December 17, 2017, www.cbsnews.com/news/lol-surprise-hot-new-toy.
18. "Toy of the Year Awards," Toy Association, www.toyassociation.org/toys/events/toy-of-the-year-awards-home.aspx.
19. Lutz Muller, "Collectibles Drive the Toy Market and Funko Is at the Wheel," March 12, 2018, *Seeking Alpha*, seekingalpha.com/article/4155536-collectibles-drive-toy-market-and-funko-is-wheel.
20. Louise Grimmer and Martin Grimmer, "Blind Bags: How Toy Makers Are Making a Fortune with Child Gambling," The Conversation, theconversation.com/blind-bags-how-toy-makers-are-making-a-fortune-with-child-gambling-127229.
21. Spring-Serenity Duvall, "Playing with Minimalism: How Parents Are Sold on High-End Toys and Childhood Simplicity," in *The Marketing of Children's Toys: Critical Perspectives on Children's Consumer Culture*, ed. Rebecca C. Haines and Nancy A. Jennings (Cham: Palgrave Macmillan, 2021).
22. Juliet B. Schor, *Born to Buy: The Commercialized Child and the New Consumer Culture* (New York: Scribner, 2004).
23. Kirk Warren Brown and Tim Kasser, "Are Psychological and Ecological Well-Being Compatible? The Role of Values, Mindfulness, and Lifestyle," *Social Indicators Research* 74, no. 2 (2005): 349–68.
24. Juliet B. Schor, *Born to Buy: The Commercialized Child and the New Consumer Culture* (New York: Scribner, 2004); Helga Dittmar et al., "The Relationship Between Materialism and Personal Well-Being: A Meta-Analysis," *Journal of Personality and Social Psychology* 107, no. 5 (2014): 879–924 doi.org/10.1037/a0037409.
25. "If Not Only GDP, What Else? Using Relational Goods to Predict the Trends of Subjective Well-Being," *International Review of Economics*

57, no. 2 (June 1, 2010): 199–213, https://doi.org/10.1007/s12232-010-0098-1.
26. Thomas Gilovich, Amit Kumar, and Lily Jampol, "A Wonderful Life: Experiential Consumption and the Pursuit of Happiness," *Journal of Consumer Psychology* 25, no. 1 (2015): 152–65, www.jstor.org/stable/26618054; Stefano Bartolini and Ennio Bilancini, "If Not Only GDP, What Else? Using Relational Goods to Predict the Trends of Subjective Well-Being," *International Review of Economics* 57, no. 2 (June 1, 2010): 199–213, doi.org/10.1007/s12232-010-0098-1.
27. Schor, *Born to Buy*.
28. Ronnel B. King and Jesus Alfonso D. Datu, "Materialism Does Not Pay: Materialistic Students Have Lower Motivation, Engagement, and Achievement," *Contemporary Educational Psychology* 49 (April 1, 2017): 289–301.
29. King and Datu, *Contemporary Educational Psychology* 49: 289–301; Agnes Nairn, Jo Ormrod, and Paul Andrew Bottomley, "Watching, Wanting and Wellbeing: Exploring the Links—a Study of 9 to 13 Year-Olds," Monograph (London: National Consumer Council, July 2007), orca.cf.ac.uk/45286.
30. Jean M. Twenge and Tim Kasser, "Generational Changes in Materialism and Work Centrality, 1976–2007: Associations with Temporal Changes in Societal Insecurity and Materialistic Role Modeling," *Personality and Social Psychology Bulletin* 39, no. 7 (May 1, 2013): 883–97, doi.org/10.1177/0146167213484586.
31. University of Sussex, "Pressure to Be Cool, Look Good Is Detrimental to Many Children," ScienceDaily, September 11, 2015, www.sciencedaily.com/releases/2015/09/150911094908.htm.
32. Marvin E. Goldberg and Gerald J. Gorn, "Some Unintended Consequences of TV Advertising to Children," *Journal of Consumer Research* 5, no. 1 (June 1, 1978): 22–29, doi.org/10.1086/208710.
33. Anna McAlister and T. Bettina Cornwell, "Children's Brand Symbolism Understanding: Links to Theory of Mind and Executive Functioning," *Psychology and Marketing* 27, no. 3 (February 11, 2010): 203–28, doi.org/10.1002/mar.20328.
34. *High Fidelity*, directed by Stephen Frears (2000; Burbank, CA: Touchstone Pictures).
35. Yalda T. Uhls, Eleni Zgourou, and Patricia M. Greenfield, "21st Century Media, Fame, and Other Future Aspirations: A National Survey of 9–15 Year Olds," *Cyberpsychology: Journal of Psychosocial Research on Cyberspace* 8, no. 4 (2014), doi.org/10.5817/CP2014-4-5; Anna Maria

Zawadzka et al., "Environmental Correlates of Adolescent Materialism: Interpersonal Role Models, Media Exposure, and Family Socio-economic Status," *Journal of Child and Family Studies*, (December 6, 2021), doi.org/10.1007/s10826-021-02180-2.

36. Alyssa Bailey, "Ariana Grande Just Released '7 Rings' and It's All About Being Rich as Hell," *Elle*, January 18, 2019, www.elle.com/culture/celebrities/a25938429/riana-grande-7-rings-lyrics-meaning.
37. Natalie Weiner, "Billboard Woman of the Year Ariana Grande: 'There's Not Much I'm Afraid of Anymore,'" *Elle*, May 12, 2018, www.billboard.com/articles/events/women-in-music/8487877/riana-grande-cover-story-billboard-women-in-music-2018.
38. "Ariana Grande Lyrics—'7 Rings,'" AZLyrics, www.azlyrics.com/lyrics/arianagrande/7rings.html.
39. Mary T. Schmich, "A Stopwatch on Shopping," *Chicago Tribune*, December 24, 1986.
40. George W. Bush, "At O'Hare, President Says 'Get on Board,'" remarks to airline employees, September 27, 2001; George W. Bush, "'Islam Is Peace' Says President," remarks at Islamic Center of Washington, DC, September 17, 2021; George W. Bush, "Statement by the President in His Address to the Nation," *PBS NewsHour*, September 11, 2001.
41. Tim Kasser, "Cultural Values and the Well-Being of Future Generations: A Cross-National Study," *Journal of Cross-Cultural Psychology* 42, no. 2 (February 21, 2011): 206–15, doi.org/10.1177/0022022110396865.
42. Tim Kasser, email message to author, April 5, 2019.

CHAPTER 5: HOW REWARDING ARE REWARDS?

Alfie Kohn, *Unconditional Parenting: Moving from Rewards and Punishments to Love and Reason* (New York: Simon and Schuster, 2006), 32.

1. *Lego City Game*, Apple Store, itunes.apple.com/us/app/lego-city-game/id1117365978?mt=8.
2. "Know Before You Load App Reviews," Children and Media Australia, childrenandmedia.org.au/app-reviews.
3. Shalom H. Schwartz, "An Overview of the Schwartz Theory of Basic Values," *Online Readings in Psychology and Culture* 2, no. 1 (December 2012), doi.org/10.9707/2307-0919.1116.
4. Schwartz, *Online Readings in Psychology and Culture*, 2.
5. Tim Kasser, *The High Price of Materialism* (Cambridge: MIT Press, 2002).

6. Kasser.
7. Kasser.
8. Frederick M.E. Grouzet et al., "The Structure of Goal Contents Across 15 Cultures," *Journal of Personality and Social Psychology* 89, no. 5 (2005): 800–16, doi.org/10.1037/0022-3514.89.5.800.
9. Tim Kasser et al., "The Relations of Maternal and Social Environments to Late Adolescents' Materialistic and Prosocial Values," *Developmental Psychology* 31, no. 6 (1995): 907–14, doi.org/10.1037/0012-1649.31.6.90; Jean M. Twenge and Tim Kasser, "Generational Changes in Materialism and Work Centrality, 1976–2007: Associations with Temporal Changes in Societal Insecurity and Materialistic Role Modeling," *Sage* 9, no. 7 (2013), doi.org/10.1177/0146167213484586.
10. Kennon M. Sheldon and Tim Kasser, "Psychological Threat and Extrinsic Goal Striving," *Motivation and Emotion* 32, no. 1 (March 1, 2008): 37–45, doi.org/10.1007/s11031-008-9081-5.
11. Marsha Richins and Lan Nguyen Chaplin, "Material Parenting: How the Use of Goods in Parenting Fosters Materialism in the Next Generation," *Journal of Consumer Research* 41, no. 6 (2015): 1333–57.
12. For an excellent review of the robust research on the impact of intrinsic and extrinsic motivators in schools, see Richard M. Ryan and Edward L. Deci, *Self-Determination Theory: Basic Psychological Needs in Motivation, Development, and Wellness* (New York: Guilford Press, 2017), especially relevant to this discussion, 319–81.
13. D.G. Singer, J.L. Singer, H. D'Agostino, and R. DeLong, "Children's Pastimes and Play in Sixteen Nations: Is Free-Play Declining?," *American Journal of Play* (Winter 2009): 283–312.
14. Edward L. Deci, "Effects of Externally Mediated Rewards on Intrinsic Motivation," *Journal of Personality and Social Psychology* 18, no. 1 (April 1971): 105–15, doi.org/10.1037/h0030644.
15. David Greene and Mark R. Lepper, "Effects of Extrinsic Rewards on Children's Subsequent Intrinsic Interest," *Child Development* 45, no. 4 (1974): 1141–45, doi.org/10.2307/1128110.
16. Alina Bradford, "Here Are All the Countries Where Pokémon Go Is Available," CNET, www.cnet.com/tech/gaming/pokemon-go-where-its-available-now-and-coming-soon.
17. "Analysis of Pokémon GO: A Success Two Decades in the Making," *Newzoo*, September 30, 2016, newzoo.com/insights/articles/analysis-pokemon-go.
18. Alysia Judge, "Pokémon GO Has Been Downloaded More Than a

Billion Times," *IGN*, August 1, 2019. www.ign.com/articles/2019/08/01/Pokémon-go-has-been-downloaded-more-than-a-billion-times.
19. "Global Mobile Games Consumer Spending 2021," Statista, September 22, 2021, www.statista.com/statistics/1179913/highest-grossing-mobile-games.
20. Susan Linn, "Confessions of a Pokémon GO Grinch: Ethical Questions About World's Most Popular App," *American Prospect*, August 3, 2016, prospect.org/article/confessions-pok%c3%a9mon-go-grinch-ethical-questions-about-world%e2%80%99s-most-popular-app.
21. "Pokémon STOP! Don't Lure Kids to Sponsor's Locations," Campaign for a Commercial-Free Childhood.
22. "Letter to Niantic," Common Sense Media, July 19, 2016, www.commonsensemedia.org/sites/default/files/uploads/kids_action/niantic_letter.7.19.16.pdf.
23. "The Mediatrician Speaks—Pokémon GO!," Center on Media and Child Health, July 22, 2019, cmch.tv/the-mediatrician-speaks-Pokémon-go.
24. Rachael Rettner, "'Pokémon GO' Catches High Praise from Health Experts," Live Science, July 12, 2016, www.livescience.com/55373-Pokémon-go-exercise.html.
25. Amy Donaldson, "Pokémon GO Manages to Disguise Exercise as Play in Ways that May Make Us Healthier Physically and Emotionally," *Deseret Morning News*, Jul 10, 2016; Matthew Byrd, "Pokémon GO and a Brief History of Accidental Exercise Through Gaming," *Den of Geek*, August 24, 2016.
26. Kate Silver, "Pokémon GO Leading to a 'Population-Level' Surge in Fitness Tracker Step Counts," *Washington Post*, July 15, 2016.
27. Rachel Bachman and Sarah E. Needleman, "Want to Exercise More? Try Screen Time; 'Pokémon GO' Got Millions of People Moving; New Research Shows Potential for Smartphone Apps and Games to Boost Physical Activity," *Wall Street Journal Online*, December 17, 2016, www.wsj.com/articles/want-to-exercise-more-try-screen-time-1481976002.
28. Katherine B. Howe et al., "Gotta Catch 'Em All! Pokémon GO and Physical Activity Among Young Adults: Difference in Differences Study," *BMJ* 355 (December 13, 2016): i6270, doi.org/10.1136/bmj.i6270.
29. Alessandro Gabbiadini, Christina Sagioglou, and Tobias Greitemeyer, "Does Pokémon GO Lead to a More Physically Active Life Style?," *Computers in Human Behavior* 84 (July 1, 2018): 258–63, doi.org/10.1016/j.chb.2018.03.005.

30. Tiffany May, "Pokémon Sleep Wants to Make Snoozing a Game Too," *New York Times*, May 29, 2019.
31. May, "Pokémon Sleep."
32. May, "Pokémon Sleep."
33. "Infographic: Sleep in the Modern Family," National Sleep Foundation, www.sleepfoundation.org/articles/infographic-sleep-modern-family.
34. Jennifer Falbe et al., "Sleep Duration, Restfulness, and Screens in the Sleep Environment," *Pediatrics* 135, no. 2 (February 2015): e367-75, doi.org/10.1542/peds.2014-2306.
35. Rob Newsom, "Relaxation Exercises for Falling Asleep," National Sleep Foundation, December 18, 2020, www.sleepfoundation.org/articles/relaxation-exercises-falling-asleep.
36. "Sleep, n.," *Oxford English Dictionary Online* (Oxford University Press), February 3, 2022, www.oed.com/view/Entry/181603.
37. May, "Pokémon Sleep."
38. "Pokémon Smile," the Pokémon Company, smile.Pokémon.com/en-us.
39. "Pokémon Smile."
40. "Pokémon Smile."
41. Collectible Mew Oreo Cookie Limited Edition Cracked, eBay, www.ebay.com/itm/324797022990?_trkparms=ispr%3D1&hash=item4b9f69430e:g:bVUAAOSwVh5hSXDn&amdata.
42. Mew Oreo Cookie Challenge!!! Try to Get to $1,000,000 Let's Go, eBay, hwww.ebay.com/itm/353684751511?hash=item5259410497:g:plcAAOSwJmZhSSrC.
43. Lulu Garcia-Navara, "Gotta Catch 'Em All: Pokémon Fans Go Crazy Over New Oreos," *NPR Weekend Edition*, September 26, 2021, www.npr.org/2021/09/26/1040756485/gotta-catch-them-all-Pokémon-fans-go-crazy-over-new-oreos.
44. Mohammad Alkilzy et al., "Improving Toothbrushing with a Smartphone App: Results of a Randomized Controlled Trial," *Caries Research* 53, no. 6 (2019): 628-35, doi.org/10.1159/000499868.
45. Fairplay, "CCFC Announces 2013 TOADY Award Nominees for Worst Toy of the Year," news release, November 21, 2013.

CHAPTER 6: THE NAGGING PROBLEM OF PESTER POWER

David Sprinkle, "Packaged Facts: 3 Mega Trends Impacting Kids Food Industry," Cision PR Newswire, March 30, 2016, www.prnewswire

.com/news-releases/packaged-facts-3-mega-trends-impacting-kids-food-industry-300242013.html.

1. Moniek Buijzen and Patti M. Valkenburg, "The Unintended Effects of Television Advertising: A Parent-Child Survey," *Communication Research* 30, no. 5 (October 2003): 483–503, doi.org/10.1177/0093650203256361; Moniek Buijzen and Patti M. Valkenburg, "The Effects of Television Advertising on Materialism, Parent–Child Conflict, and Unhappiness: A Review of Research," *Journal of Applied Developmental Psychology* 24, no. 4 (September 1, 2003): 437–56, doi.org/10.1016/S0193-3973(03)00072-8.
2. See Kalpesh B. Prajapati, Trivedi Payal, and Sneha Advani, "A Study on Pester Power's Impact: 'Identify Pester Power's Impact on the Purchase Decision of Guardians with Specific Focus on Snacks Category,'" *Asian Journal of Research in Business Economics and Management* 4, no. 2 (2014): 143–49; Eileen Bridges and Richard A. Briesch, "The 'Nag Factor' and Children's Product Categories," *International Journal of Advertising* 25, no. 2 (May 2006): 157–87.
3. Holly K.M. Henry and Dina L.G. Borzekowski, "The Nag Factor," *Journal of Children and Media* 5, no. 3 (2011): 298–317; Buijzen and Valkenburg, *Communication Research* 30: 483–503.
4. Laura McDermott et al., "International Food Advertising, Pester Power and Its Effects," *International Journal of Advertising* 25, no. 4 (2006): 513–40.
5. Widmeyer Communications, *Kids and Commercialism Final Results*, report for the Center for a New American Dream, December 13, 2002, 7, newdream.org/resources/poll-kids-and-commercialism-2002.
6. Henry and Borzekowski, *Journal of Children and Media* 5: 298–317.
7. James U. McNeal, *On Becoming a Consumer: Development of Consumer Behavior Patterns in Childhood* (London: Elsevier, 2007), 240–41.
8. Elliott Haworth, "Banning Junk Food Ads Won't Stem the Flow of Fat Kids," *CITY A.M.*, December 12, 2016, www.cityam.com/255380/banning-junk-food-ads-wont-stem-flow-fat-kids.
9. Elliott Haworth, "#Advert #Kids; Holly Worrollo Teaches Elliott Haworth How to Target Young People," *CITY A.M.*, August 7, 2017, advance-lexis-com.ezp-prod1.hul.harvard.edu/api/document?collection=news&id=urn:contentItem:5P67-FWX1-DY9P-N1F1-00000-00&context=1516831.
10. Haworth, "#Advert #Kids."
11. UN General Assembly, Resolution 44/25, Convention on the Rights

of the Child, A/RES/44/25 (November 20, 1989), undocs.org/en/A/RES/44/25.
12. Larry Kramer, "Kids Advertising Hearings to Open," *Washington Post*, February 28, 1979.
13. "Stop Press," *Marketing*, November 25, 1993, advance-lexis-com.ezp-prod1.hul.harvard.edu/api/document?collection=news&id=urn:contentItem:419K-5BS0-00X8-J4PV-00000-00&context=1516831.
14. Western Media International, "The Fine Art of Whining: Why Nagging Is a Kid's Best Friend," *Business Wire*, August 11, 1998.
15. Amy Frazier, "Market Research: The Old Nagging Game Can Pay Off for Marketers," *Selling to Kids* 3, no. 8 (April 15, 1998).
16. Frazier, *Selling to Kids* 3.
17. "From Nag to the Bag," *Brandweek*, New York, Apr 13, 1998.
18. Elena Morales, "The Nag Factor: Measuring Children's Influence," *Admap*, March 2000, 35–37.
19. Megan Sanks, "Your Complete Guide to Everything Owned by Viacom," Zacks, July 10, 2017, www.zacks.com/stock/news/267014/your-complete-guide-to-everything-owned-by-viacom.
20. "Kidfluence: How Kids Influence Buying Behavior," Viacom, March 28, 2018, www.paramount.com/news/audience-insights/kidfluence-kids-influence-buying-behavior. Viacom relies on Martin Lindstrom's widely cited 2004 estimate of annual purchasing power in Martin Lindstrom, "Branding Is No Longer Child's Play!" *Journal of Consumer Marketing* 21, no. 3 (May 1, 2004): 175–82, doi.org/10.1108/07363760410534722.
21. "Kidfluence: How Kids Influence Buying Behavior," Viacom, March 28, 2018, www.paramount.com/news/audience-insights/kidfluence-kids-influence-buying-behavior.
22. Western Media International, *Business Wire*, 1998.
23. Lucy Hughes, interview by Mark Achbar, *The Corporation*, directed by Mark Achbar (New York: Big Picture Media Corporation, 2005), DVD.
24. Achbar, *The Corporation*.
25. See *Consuming Kids, The Commercialization of Childhood* (New York: Media Education Foundation, 2008), eVideo; Margo G. Wootan, *Pestering Parents: How Food Companies Market Obesity to Children* (Washington, DC: Center for Science in the Public Interest, November 1, 2003), www.cspinet.org/new/200311101.html. ; Juliet B. Schor, *Born to Buy: The Commercialized Child and the New Consumer Culture* (New York: Scribner, 2004); Susan Linn, *Consuming Kids: The Hostile Takeover of Childhood* (New York: New Press, 2005).

26. Gary Ruskin, quoted in *Consuming Kids, The Commercialization of Childhood* (New York: Media Education Foundation, 2008), eVideo.
27. Peter Waterman, quoted in Diane Summers, "Harnessing Pester Power," *Financial Times*, December 23, 1993, 8.
28. European Commission, "Consumers: New EU Rules Crackdown on Misleading Advertising and Aggressive Sales Practices," news release, December 12, 2007, europa.eu/rapid/press-release_IP-07-1915_en.htm.
29. Children's Advertising Review Unit, "Self-Regulatory Program for Children's Advertising," fkks.com/uploads/news/6.30.11_CARU_Ad_Guidelines.pdf.
30. Nintendo, "Why We Need Wii U," 2013, www.ispot.tv/ad/75j9/nintendo-wii-u-why-we-need-wii-u.
31. Mahsa-Sadat Taghavi and Alireza Seyedsalehi, "The Effect of Packaging and Brand on Children's and Parents' Purchasing Decisions," *British Food Journal* 117, no. 8 (2015): 2017–38, doi.org/10.1108/BFJ-07-2014-0260.
32. Taghavi and Seyedsalehi, "The Effect of Packaging and Brand on Children's and Parents' Purchasing Decisions," 2017–38.
33. Monica Chaudhary, "Pint-Size Powerhouses: A Qualitative Study of Children's Role in Family Decision-Making," *Young Consumers* 19, no. 4 (2018): 345–57, doi.org/10.1108/YC-04-2018-00801.
34. Rituparna Bhattacharyya and Sangita Kohli, "Target Marketing to Children: The Ethical Aspect" (Proceedings of the International Marketing Conference on Marketing and Society, Indian Institute of Management, Kozhikode, India, April 2007), 69–74.
35. Lindstrom, *Journal of Consumer Marketing* 21:175–82.
36. ŠKODA Auto India, "ŠKODA Launches Rapid Onyx in India; Unveils Integrated Ad Campaign to Promote the New Model," news release, *Adobo Magazine*.
37. Zainub Ali, "The Payoff in Marketing to Kids," *Slogan* 17, no. 1 (January 2012), 24.
38. Callie Watson, "Parents Resisting Pester Power," *The Advisor* (June 3, 2014), 7.
39. "Press Release: A Third of Parents in Scotland Say 'Pester Power' Regularly Gets the Better of Them," *Money Advice Service*," November 27, 2018, www.moneyadviceservice.org.uk/en/corporate/press-release—a-third-of-parents-in-scotland-say-pester-power-regularly-gets-the-better-of-them.

40. Helena Pozniak, "*Fortnite* Fury: The Extremes Parents Are Going To, to Get Their Children Off This Addictive Game," *Telegraph*, September 27, 2018.
41. Pozniak, "Fortnite Fury." See also Jef Feeley and Christopher Palmeri, "Kids Are So Addicted to *Fortnite* They're Being Sent to Gamer Rehab," *Bloomberg*, November 27, 2018, www.bloomberg.com/news/articles/2018-11-27/fortnite-addiction-prompts-parents-to-turn-to-video-game-rehab; Elizabeth Matsangou "How Fortnite Became the Most Successful Free-to-Play Game Ever," *New Economy*, November 14, 2018. www.theneweconomy.com/business/how-fortnite-became-the-most-successful-free-to-play-game-ever.
42. See Nir Eyal, *Hooked: How to Build Habit-Forming Products* (New York: Portfolio/Penguin, 2014); Adam Alter, *Irresistible: The Rise of Addictive Technology and the Business of Keeping Us Hooked* (New York: Penguin, 2017).
43. Marisa Meyer et al., "Advertising in Young Children's Apps: A Content Analysis," *Journal of Developmental and Behavioral Pediatrics* 40, no. 1 (2019): 32–39.
44. Nellie Bowles, "Your Kid's Apps Are Crammed with Ads," *New York Times*, October 30, 2018.
45. Charles K. Atkin, "Observation of Parent-Child Interaction in Supermarket Decision-Making," *Journal of Marketing* 42, no. 4 (1978): 41–45, doi.org/10.2307/1250084.
46. James U. McNeal, *Kids as Customers: A Handbook of Marketing to Children* (Lanham, MD: Lexington Books, 1992), 79.
47. Richard Fry, "Millennials Overtake Baby Boomers as America's Largest Generation," Pew Research Center, April 25, 2016, www.pewresearch.org/fact-tank/2016/04/25/millennials-overtake-baby-boomers.
48. "Brand Research, Strategy, and Innovation," The Family Room, familyroomllc.com/home-page.
49. The quotes from George Carey in this paragraph and in the following section are from "The Day the Universe Changed," a keynote speech I heard him give on February 12, 2018, at the Kidscreen Summit in Miami, Florida, summit.kidscreen.com/2018/sessions/68723/keynote thedaythe.
50. 2014 census data quoted in Konrad Mugglestone, "Finding Time: Millennial Parents, Poverty and Rising Costs," Young Invincibles, May 2015, younginvincibles.org/wp-content/uploads/2017/04/Finding-Time-Apr29.2015-Final.pdf.

51. Richard Fry, "5 Facts About Millennial Households," Pew Research Center, September 6, 2017, www.pewresearch.org/fact-tank/2017/09/06/5-facts-about-millennial-households.
52. Ryan Jenkins, "This Is Why Millennials Care So Much About Work-Life Balance," Inc., January 8, 2018, www.inc.com/ryan-jenkins/this-is-what-millennials-value-most-in-a-job-why.html.
53. Ace Casimiro, "Employee Engagement: Out of Office, Seldom Away from Work," *Randstad*, rlc.randstadusa.com/for-business/learning-center/workforce-management/employee-engagement-out-of-office-seldom-away-from-work.
54. Steve McClellan, "Moms Relinquish Role as Family Decision Maker," May 11, 2017, Media Post, www.mediapost.com/publications/article/300952/moms-relinquish-role-as-family-decision-maker.html.
55. Henry and Borzekowski, *Journal of Children and Media* 5: 298–317.

Chapter 7: Divisive Devices

Jenny Radesky, "What Happens When We Turn to Smartphones to Escape Our Kids?," *Pacific Standard*, April 22, 2019, psmag.com/ideas/what-happens-when-parents-cant-put-their-phones-down.

1. Jenny S. Radesky et al., "Patterns of Mobile Device Use by Caregivers and Children During Meals in Fast Food Restaurants," *Pediatrics* 133, no. 4 (March 2014): 843–49, doi.org/10.1542/peds.2013-3703; Alexis Hiniker et al., "Texting While Parenting: How Adults Use Mobile Phones While Caring for Children at the Playground" (CHI '15: Proceedings of the 33rd Annual ACM Conference on Human Factors in Computing Systems, April 2015), 727–36.
2. Hiniker et al., "Texting While Parenting," 727–36.
3. Radesky et al., *Pediatrics* 133: 843–49.
4. Hiniker et al., "Texting While Parenting," 727–36.
5. Brandon T. McDaniel and Jenny S. Radesky, "Technoference: Parent Distraction with Technology and Associations with Child Behavior Problems," *Child Development* 89, no. 1 (2018): 100–109, doi.org/10.1111/cdev.12822.
6. McDaniel and Radesky, *Child Development* 89: 100–109.
7. Adrienne LaFrance, "The Golden Age of Reading the News," *The Atlantic*, May 6, 2016.
8. Tiffany G. Munzer et al., "Tablets, Toddlers and Tantrums: The

Immediate Effects of Tablet Device Play," *Acta Paediatrica* 110, no. 1 (January 1, 2021): 255–56.

9. Alexis Hiniker et al., "Let's Play: Digital and Analog Play Between Preschoolers and Parents" (CHI '18: Proceedings of the 2018 CHI Conference on Human Factors in Computing Systems, April 2018), 1–13, doi.org/10.1145/3173574.3174233.

10. Kathy Hirsh-Pasek et al., "Putting Education in 'Educational' Apps: Lessons from the Science of Learning," *Psychological Science in the Public Interest* 16, no. 1 (2015): 3–34; Meyer et al., *Journal of Developmental and Behavioral Pediatrics* 40: 32–39.

11. Douglas A. Gentile et al., "Pathological Video Game Use Among Youths: A Two-Year Longitudinal Study," *Pediatrics* 127, no. 2 (February 2011): e319-29; Clifford J. Sussman et al., "Internet and Video Game Addictions: Diagnosis, Epidemiology, and Neurobiology," *Child and Adolescent Psychiatric Clinics of North America* 27, no. 2 (April 2018): 301–26.

12. Betsy Morris, "How Fortnite Triggered an Unwinnable War Between Parents and Their Boys; the Last-Man-Standing Videogame Has Grabbed onto American Boyhood, Pushing Aside Other Pastimes and Hobbies and Transforming Family Dynamics," *Wall Street Journal Online*, December 21, 2018.

13. Nick Bilton, "Steve Jobs Was a Low-Tech Parent," *New York Times*, September 10, 2014.

14. Bianca Bosker, "Tristan Harris Believes Silicon Valley Is Addicting Us to Our Phones. He's Determined to Make It Stop," *Atlantic Monthly* 318, no. 4 (November 2016): 56–58, 60, 62, 64–65, search.proquest.com/docview/1858228137/abstract/FBF7AA159CE64354PQ/1.

15. H. Tankovska, "Facebook's Annual Revenue from 2009 to 2020," Statista (Facebook Annual Report 2021, February 5, 2021), 66, www.statista.com/statistics/268604/annual-revenue-of-facebook.

16. Jaron Lanier quoted in Jemima Kiss, "'I Was on Instagram. The Baby Fell Down the Stairs': Is Your Phone Use Harming Your Child?," *The Guardian*, December 7, 2018.

17. Victoria Rideout and Michael Robb, *The Common Sense Census: Media Use by Tweens and Teens* (San Francisco: Common Sense Media, 2019).

18. Doreen Carvajal, "A Way to Calm a Fussy Baby: 'Sesame Street' by Cellphone," *New York Times*, April 18, 2005.

19. Michele M. Melendez, "Calling on Kids; Cell Phone Industry Aims at Youngest Consumers," *Grand Rapids Press*, July 3, 2005.

20. Ken Heyer, quoted in Doreen Carvajal, "A Way to Calm a Fussy Baby: 'Sesame Street' by Cellphone," *New York Times*, April 18, 2005.
21. Sean Dennis Cashman, *America in the Twenties and Thirties: The Olympian Age of Franklin Delano Roosevelt* (New York: New York University Press, 1989), 63.
22. Cashman, *America in the Twenties and Thirties*, 329.
23. Robert Putnam, *Bowling Alone: The Collapse and Revival of American Community* (New York: Simon and Schuster, 2000).
24. Tiffany G. Munzer et al., "Parent-Toddler Social Reciprocity During Reading from Electronic Tablets vs. Print Books," *JAMA Pediatrics* 173, no. 11 (2019): 1076–83, www.ncbi.nlm.nih.gov/pmc/articles/PMC6777236; Jennifer M. Zosh et al., "Talking Shape: Parental Language with Electronic Versus Traditional Shape Sorters," *Mind, Brain, and Education* 9, no. 3 (September 2015): 136–44, doi.org/10.1111/mbe.12082.
25. Tiffany Munzer et al., "Differences in Parent-Toddler Interactions with Electronic Versus Print Books," *Pediatrics* 143, no. 4 (2019): e20182012, doi-org.ezp-prod1.hul.harvard.edu/10.1542/peds.2018-2012; Munzer et al., "Parent-Toddler Social Reciprocity During Reading from Electronic Tablets vs. Print Books."
26. Sherry Turkle, *Reclaiming Conversation: The Power of Talk in the Digital Age* (New York: Penguin Press, 2015), 107.
27. "Best 5 Smart Toasters for You to Buy in 2020 Reviews," toasteraddict.com/smart-toaster; Kimiko de Freytas-Tamura, "Maker of 'Smart' Vibrators Settles Data Collection Lawsuit for $3.75 Million," *New York Times*, March 14, 2017.
28. "Part 1: Digital Assistants, the Internet of Things, and the Future of Search," Ignite Technologies, October 14, 2016, placeable.ignitetech.com/blog/post/part-1-digital-assistants-internet-things-future-search.
29. Jason England, "Amazon Dominates the Smart Speaker Market with a Nearly 70% Share in 2019," Android Central, February 11, 2020, www.androidcentral.com/amazon-dominates-smart-speaker-market-nearly-70%-share-2019.
30. Amazon, "Echo Dot (3rd Gen), Smart Speaker with Alexa, Charcoal," Amazon Devices, www.amazon.com/Echo-.Dot/dp/B07FZ8S74R/ref=sr_1_1?dchild=1&keywords=echo&qid=1617893811&sr=8-1.
31. "Create Alexa Skills Kit, Amazon Alexa Voice Development," Amazon, Alexa, developer.amazon.com/en-US/alexa/alexa-skills-kit.html.
32. Daniel Etherington, "Amazon Echo is a $199 Connected Speaker Packing an Always-On Siri-Style Assistant," *TechCrunch*, November 6, 2014, techcrunch.com/2014/11/06/amazon-echo.

33. Sarah Perez, "Hands On with the Echo Dots Kids Edition," *TechCrunch*, June 28, 2018, techcrunch.com/2018/06/28/hands-on-with-the-echo-dots-kids-edition.
34. Amazon, "Introducing Fire HD Kids Edition—The Kids Tablet that Has It All, Including the First-Ever 2-Year Worry-Free Guarantee," news release, September 17, 2014, press.aboutamazon.com/news-releases/news-release-details/introducing-fire-hd-kids-edition-kids-tablet-has-it-all.
35. Igor Bonifacic, "Amazon Announces Its First-Ever Kindle for Kids," *Engadget*, October 7, 2019; Sarah Perez, "Amazon Introduces a Way for Teens to Independently Shop Its Site, Following Parents' Approval," *TechCrunch*, October 11, 2017.
36. Todd Haselton, "Amazon Targets Kids with a Candy-Colored Echo and a Version of Alexa that Rewards Politeness," *CNBC*, April 25, 2018, www.cnbc.com/2018/04/25/amazon-echo-kids-edition-alexa-for-kids.html.
37. "Echo Dot Kids Edition Violates COPPA," request for investigation of Amazon's violations of the Children's Online Privacy Protection Act (COPPA) for the safety and privacy of American children, May 9, 2019, www.echokidsprivacy.com.
38. "Echo Dot Kids Edition Violates COPPA," iii.
39. "Echo Dot Kids Edition Violates COPPA," 31–32.
40. "Echo Dot Kids Edition Violates COPPA," iv.
41. "Echo Dot Kids Edition Violates COPPA," 24.
42. Josh Golin, personal communication to author, April 8, 2021; Angela Campbell, personal communication to author, November 8, 2021.
43. Amazon, "All-New Echo Dot (4th Gen) Kids Edition, Designed for kids, with parental controls, Tiger," Amazon Devices, www.amazon.com/Echo-Dot-4th-Gen-Kids/dp/B084J4QQK1.
44. Amazon, "All-New Echo Dot (4th Gen) Kids Edition."
45. Amazon, "All-New Echo Dot (4th Gen) Kids Edition."
46. Amazon, "All-New Echo Dot (4th Gen) Kids Edition."
47. Amazon, "All-New Echo Dot (4th Gen) Kids Edition."
48. Emily DeJeu, "6 Ways the Amazon Echo Will Transform Your Child's Bedtime Routine," The Baby Sleep Site, August 5, 2020, www.babysleepsite.com/sleep-training/amazon-echo-bedtime-routine.
49. Ryan Tuchow, "What's Now vs. What's next: The Future of Kids Tech," *Kidscreen*, March 15, 2022, kidscreen.com/2022/03/15/whats-now-vs-whats-next-the-future-of-kids-tech.
50. For information about how to play Poohsticks see "In Which Pooh

Invents a Game and Eeyore Joins In," *The House at Pooh Corner* (Boston: E.P. Dutton, 1928), 92–108.

51. Amazon, "All-New Echo Dot (4th Gen) Kids Edition + Echo Glow, Panda," Amazon Devices, www.amazon.com/Echo-Designed-parental-controls-Panda/dp/B084J4MJCK.
52. The latest Echo Dot for kids is posted as suitable for kids ages three and up, as seen here: www.amazon.com/Echo-Dot-4th-Gen-Kids/dp/B084J4QQK1?th=1.
53. parents.amazon.com/settings/smartfilter/KTYWP4Q7FQ.
54. "American Girl Dolls Wiki: Julie Albright (Doll)," American Girl Wiki, Fall 2014, americangirl.fandom.com/wiki/Julie_Albright_(doll).
55. "Julie," doll, American Girl, Shop, Historical Characters, www.americangirl.com/shop/c/julie.
56. *An American Girl Story*, directed by Sasie Sealy, season 103 (Amazon Studios, 2017), www.amazon.com/American-Girl-Story-Julie-Balance/dp/B086HXZ4Q9/ref=sr_1_5?dchild=1&keywords=julie+albright+movie&qid=1616692782&s=movies-tv&sr=1-5.
57. American Girl, "And the Tiara Goes To: A Julie Albright Movie, Julie Albright, @American Girl," YouTube video, www.youtube.com/watch?v=F6wp23uh9o0.
58. Adrienne Jeffries and Leon Yin, "Google's Top Search Result? Surprise! It's Google," Mark Up, Google the Giant, themarkup.org/google-the-giant/2020/07/28/google-search-results-prioritize-google-products-over-competitors.
59. Lauren Johnson, "Amazon Is Ramping Up Its Pitch for Audio Ads After Long Promising Alexa Would Be Ad-Free," *Business Insider*, www.businessinsider.com/amazon-is-pitching-advertisers-more-on-alexa-audio-ads-2020-2.
60. Rules about host-selling are covered in "The Public and Broadcasting," Federal Communications Commission, December 7, 2015, www.fcc.gov/media/radio/public-and-broadcasting.
61. Mila Slesar, "Amazon Alexa: How to Leverage the Benefits for Your Brand," Alternative Spaces, alternative-spaces.com/blog/amazon-alexa-how-to-leverage-the-benefits-for-your-brand.
62. Tharin White, "Disney Adds Amazon Echo 'Frozen' and 'Star Wars' Voice Skills for Kids+," *Attractions Magazine*, February 3, 2021, attractionsmagazine.com/disney-adds-amazon-echo-frozen-star-wars-voice-skills-kids.
63. Morten L. Kringelbach et al., "On Cuteness: Unlocking the Parental

Brain and Beyond," *Trends in Cognitive Sciences* 20, no. 7 (July 1, 2016): 545–58, doi.org/10.1016/j.tics.2016.05.003.

64. Junya Nakanishi, Jun Baba, and Itaru Kuramoto "How to Enhance Social Robots' Heartwarming Interaction in Service Encounters" (Proceedings of the 7th International Conference on Human-Agent Interaction, September, 2019), 297–99, dl-acm-org.ezp-prod1.hul.harvard.edu/doi/epdf/10.1145/3349537.3352798; Also, for a description of how designers thought about a specific robot's physical structure that engages people, see Alexander Reben and Joseph Paradiso, "A Mobile Interactive Robot for Gathering Structured Social Video" (Proceedings of the 19th Association for Computing Machinery, International Conference on Multimedia, 2011), 917–20, doi.org.ezp-prod1.hul.harvard.edu/10.1145/2072298.2071902.

65. Sherry Turkle quoted in James Vlahos, "Barbie Wants to Get to Know Your Child," *New York Times*, September 16, 2015.

66. See Mattel's list of Hello Barbie's possible responses at "What Does Hello Barbie Say?," hellobarbiefaq.mattel.com/what-does-hello-barbie-say.

67. For an in-depth exploration of children and their relationships with robotic-like toys, such as Furby, see Sherry Turkle, *Alone Together: Why We Expect More from Technology and Less from Each Other* (New York: Basic Books, 2012), particularly pages 27–52. Also, Peter H. Kahn Jr. et al., "'Robovie, You'll Have to Go into the Closet Now': Children's Social and Moral Relationships with a Humanoid Robot," *Developmental Psychology, Interactive Media and Human Development* 48, no. 2 (2012): 303–14, doi.org/10.1037/a0027033.

68. V.I. Kraak and M. Story, "Influence of Food Companies' Brand Mascots and Entertainment Companies' Cartoon Media Characters on Children's Diet and Health: A Systematic Review and Research Needs," *Obesity Reviews* 16, no. 2 (2015): 107–26, doi.org/10.1111/obr.12237.

69. Sandra L. Calvert et al., "Young Children's Mathematical Learning from Intelligent Characters," *Child Development* 91, no. 5 (2020): 1491–1508, doi.org/https://doi.org/10.1111/cdev.13341.

70. Calvert et al., "Young Children's Mathematical Learning from Intelligent Characters."

71. Larry D. Woodard, "Dora the (Marketing) Explorer," *ABC News*, April 20, 2010.

Chapter 8: Bias for Sale

D. Fox Harrell, quoted in Elisabeth Soep, "Chimerical Avatars and Other Identity Experiments from Prof. Fox Harrell," *Boing Boing*, April 19, 2010, boingboing.net/2010/04/19/chimerical-avatars-a.html.

1. Federal Trade Commission, "Ads Touting 'Your Baby Can Read' Were Deceptive, FTC Complaint Alleges," August 28, 2012, www.ftc.gov/news-events/press-releases/2012/08/ads-touting-your-baby-can-read-were-deceptive-ftc-complaint.
2. Monica Potts, "It's an Ad World After All: Is It Legal for a Company to Take Out Internet Ads on Your Name After You've Filed a Complaint Against It? Apparently So," *American Prospect* 22, no. 9, (2011): 42, prospect.org/environment/ad-world.
3. Latanya Sweeney, "Discrimination in Online Ad Delivery," *ACM Queue* (a journal published by the Association of Computer Machinery) 11, no. 3 (2013), queue.acm.org/detail.cfm?id=2460278&doi=10.1145%2F2460276.2460278.
4. "1 Second, Internet Live Stats," Internetlivestats.com, www.internetlivestats.com/one-second/#google-band.
5. "Dossier on Online Search Usage," Statista, 2020, 61, www.statista.com/study/15884/search-engine-usage-statista-dossier. The data reported is global.
6. "Dossier on Online Search Usage," 3.
7. Kristen Purcell, Joanna Brenner, and Lee Rainie, "Main Findings," Pew Research Center, Internet and Technology, March 9, 2012, www.pewresearch.org/internet/2012/03/09/main-findings-11.
8. Ajit Pai, "What I Hope to Learn from the Tech Giants," *Medium*, Personal Page of the Chairman of the Federal Communications Commission, September 4, 2018, ajitvpai.medium.com/what-i-hope-to-learn-from-the-tech-giants-6f35ce69dcd9.
9. Dylan Curran, "Are You Ready? Here Is All the Data Facebook and Google Have on You," *Guardian*, March 30, 2018.
10. For examples of how much SEO companies can charge, see "How Much Does SEO Cost in 2021?" WebFX, www.webfx.com/internet-marketing/how-much-does-seo-cost.html; or Justin Smith, "SEO Pricing, How Much Do SEO Services Cost in 2021?," July 21, 2021, OuterBox, www.outerboxdesign.com/search-marketing/search-engine-optimization/seo-pricing-costs.
11. Megan Graham and Jennifer Elias, "How Google's $150 Billion Advertising Business Works," CNBC, May 18, 2021, www.cnbc.com/2021

/05/18/how-does-google-make-money-advertising-business-break down-.html.
12. Kirsten Grind et al., "How Google Interferes with Its Search Algorithms and Changes Your Results," *Wall Street Journal*, November 15, 2019, www.wsj.com/articles/how-google-interferes-with-its-search-algorithms-and-changes-your-results-11573823753?reflink =desktopwebshare_permalink.
13. Safiya Umoja Noble, *Algorithms of Oppression: How Search Engines Reinforce Racism* (New York: New York University Press), 28.
14. Noble, 19, 67.
15. See Ruha Benjamin, *Race After Technology: Abolitionist Tools for the New Jim Code* (Cambridge: Polity Press, 2019), 93–95.
16. Rebecca Hersher, "What Happened When Dylann Roof Asked Google for Information About Race?," NPR, The Two-Way, January 20, 2017.
17. Leon Yin and Aaron Sankin, "Google Ad Portal Equated 'Black Girls' with Porn," *The Markup*, July 23, 2020, themarkup.org/google-the -giant/2020/07/23/google-advertising-keywords-black-girls.
18. Hersher, "Dylann Roof."
19. Jessica Guynn, "'Three Black Teenagers' Google Search Sparks Outrage," *USA Today*, June 9, 2016, www.usatoday.com/story/tech/news /2016/06/09/google-image-search-three-black-teenagers-three-white -teenagers/85648838.
20. Sundar Pichai, "Our Commitments to Racial Equity," A Message from Our CEO, Google, June 17, 2020, blog.google/inside-google/company -announcements/commitments-racial-equity.
21. Meredith Broussard, "When Algorithms Do the Grading," *New York Times*, September 9, 2020.
22. Noble, *Algorithms of Oppression*, 82.
23. Eric Hall Schwarz, "Alexa Introduces Voice Profiles for Kids and New AI Reading Tutor," Voicebot.ai, June 29, 2021, voicebot.ai/2021/06 /29/alexa-introduces-voice-profiles-for-kids-and-new-ai-reading-tutor.
24. Monica Anderson et al., "Public Attitudes Toward Political Engagement on Social Media," Pew Research Center, Internet & Technology, Activism in the Social Media Age, July 11, 2018, www.pewresearch.org /internet/2018/07/11/public-attitudes-toward-political-engagement-on -social-media.
25. Linette Lopez, "Facebook's Algorithm Is a Sociopath, and Facebook Management Is Too Greedy to Stop It," *Business Insider*, May 28, 2020, www.businessinsider.com/facebook-algorithm-sociopath-man agement-too-greedy-to-stop-it-2020-5.

26. Davey Alba, "'All of It Is Toxic': A Surge in Protest Misinformation," *New York Times,* June 2, 2020; Davey Alba, "Misinformation About George Floyd Protests Surges on Social Media," *New York Times*, June 1, 2020.
27. See for instance Mary Aiken, "The Kids Who Lie About Their Age to Join Facebook," *The Atlantic*, August 30, 2016; "U.S. Popular Entertainment for Kids and Teens 2020," Statista, January 26, 2021, www.statista.com/statistics/1155926/popular-entertainment-us-kids-teens-parents; "Leading Social Media Apps Children UK 2020," Statista, July 21, 2021, www.statista.com/statistics/1124966/leading-social-media-apps-children-uk; "U.S. Young Kids Top Social Networks 2020," Statista, January 28, 2021, www.statista.com/statistics/1150621/most-popular-social-networks-children-us-age-group.
28. Michael B. Robb, *News and America's Kids: How Young People Perceive and Are Impacted by the News* (San Francisco: Common Sense Media, 2017), 5.
29. Tate-Ryan Mosely, "How Digital Beauty Filters Perpetuate Colorism," *MIT Technology Review*, August 15, 2021, www.technologyreview.com/2021/08/15/1031804/digital-beauty-filters-photoshop-photo-editing-colorism-racism.
30. Alex Hern, "Student Proves Twitter Algorithm 'Bias' Toward Lighter, Slimmer, Younger Faces," *The Guardian*, August 10, 2021.
31. Christopher P. Barton and Kyle Somerville, *Historical Racialized Toys in the United States (Guides to Historical Artifacts)* (New York: Routledge, 2016), 17–18.
32. Kirsten Weir, "Raising Anti-Racist Children. Psychologists Are Studying the Processes by Which Young Children Learn About Race—and How to Prevent Prejudice from Taking Root," *American Psychological Association*, 52, no. 4 (June 2021), www.apa.org/monitor/2021/06/anti-racist-children.
33. Weir, *American Psychological Association.*
34. Emily R. Aguiló-Pérez, "Commodifying Culture: Mattel's and Disney's Marketing Approaches to 'Latinx' Toys and Media," in *The Marketing of Children's Toys*, ed. Rebecca C. Hains and Nancy A. Jennings (Cham: Palgrave Macmillan, 2021), 143–63, doi.org/10.1007/978-3-030-62881-9_8.
35. Mail Foreign Service, "Barbie Launch First Black Doll That Is NOT Just a Painted Version of White Doll," *Mail Online,* October 12, 2009, www.dailymail.co.uk/news/article-1219257/Barbie-launch-black-doll-look-like-real-people-having-fuller-features.html.

36. Barton and Somerville, *Historical Racialized Toys*, 26.
37. See Liz Mineo, "The Scapegoating of Asian Americans," *Harvard Gazette* (blog), March 24, 2021, news.harvard.edu/gazette/story/2021/03/a-long-history-of-bigotry-against-asian-americans.
38. *Encyclopedia Britannica Online*, s.v. "Ku Klux Klan," www.britannica.com/topic/Ku-Klux-Klan.
39. *Encyclopedia Britannica Online*, s.v. "Tulsa Race Massacre of 1921," www.britannica.com/event/Tulsa-race-massacre-of-1921.
40. *Encyclopedia Britannica Online*, s.v. "Plessy v. Ferguson," by Brian Duignan, www.britannica.com/event/Plessy-v-Ferguson-1896.
41. Barton and Somerville, *Historical Racialized Toys*, 64.
42. Barton and Somerville, *Historical Racialized Toys*, 41.
43. Caitlin O'Kane, "Disney Adding Disclaimer About Racist Stereotypes to Some Old Movies," CBS News, October 19, 2020, www.cbsnews.com/news/disney-disclaimer-racist-stereotypes-old-movies.
44. Jessica Sullivan et al., "Adults Delay Conversations About Race Because They Underestimate Children's Processing of Race," *Journal of Experimental Psychology: General* 150, no. 2 (February 2021): 395–400, doi.org/10.1037/xge0000851.
45. Susan Linn, "*Consuming Kids: The Hostile Takeover of Childhood* (New York: New Press, 2004).

Chapter 9: Branded Learning

Marion Nestle, "School Food, Public Policy, and Strategies for Change," in *Food Politics*, ed. S. Robert and M. Weaver-Hightower (New York: Peter Lang, 2011), 143–46.

1. "JA BizTown and JA Finance Park, Junior Achievement of Greater St. Louis," www.juniorachievement.org/web/ja-gstlouis/ja-finance-park-ja-biztown.
2. JA Finance Park, "Funding Opportunities," Junior Achievement of Southern California, jasocal.org/wp-content/uploads/2019/10/FP_Funding-Opportunities_FY19.pdf.
3. JA Finance Park, "Funding Opportunities."
4. "Learn and Discover," Disneynature, Educational Materials, nature.disney.com/educators-guides.
5. Jie Jenny Zou, "Oil's Pipeline to America's Schools," Center for Public Integrity, June 15, 2017, apps.publicintegrity.org/oil-education.
6. Beth Mole, "Juul Gave Presentations in Schools to Kids, and the FDA

Is Fuming," *Ars Technica*, October 9, 2019, arstechnica.com/science/2019/09/juul-gave-presentations-in-schools-to-kids-and-the-fda-is-fuming.
7. Pizza Hut, "The BOOK IT! Program," www.bookitprogram.com.
8. "Oswego McDonald's Hosts McTeacher's Night for Boulder Hill Elementary School," *Daily Herald*, November 17, 2017, www.dailyherald.com/article/20171116/submitted/171119248.
9. Alex Molnar, *Giving Kids the Business: The Commercialization of America's Schools* (Boulder, CO: Westview Press, 1996), 39.
10. Inger L. Stole, "Advertisers in the Classroom: A Historical Perspective" (Paper presented at the Association for Consumer Research Annual Conference, Columbus, Ohio, 1999).
11. Inger L. Stole and Rebecca Livesay, "Consumer Activism, Commercialism, and Curriculum Choices: Advertising in Schools in the 1930s," *Journal of American Culture* 30, no. 1 (March 2007): 68–80, doi.org/10.1111/j.1542-734X.2007.00465.x.
12. Stole and Livesay, *Journal of American Culture* 30: 68–80.
13. Elizabeth A. Fones-Wolf, *Selling Free Enterprise: The Business Assault on Labor and Liberalism 1945–60* (Urbana: University of Illinois Press, 1994), 204. She cites *New York Times*, January 4, 1959, and others.
14. *National Commission on Excellence in Education, A Nation at Risk: The Imperative for Educational Reform* (Washington, DC: The National Commission on Excellence in Education, 1983).
15. Alex Molnar, *Giving Kids the Business: The Commercialization of America's Schools* (Boulder, CO: Westview Press, 1996), 1.
16. Richard Rothstein, "When States Spend More," *American Prospect* 9, no. 36 (January/February, 1998): 72–79.
17. Ivy Morgan and Ary Amerikaner, "Funding Gaps 2018," Education Trust, edtrust.org/resource/funding-gaps-2018.
18. Robert Hanna, Max Marchitello, and Catherine Brown, "Comparable but Unequal," Center for American Progress, March 11, 2015, americanprogress.org/issues/education-k-12/reports/2015/03/11/107985/comparable-but-unequal.
19. For a good discussion, see Alex Molnar and Faith Boninger, *Sold Out: How Marketing in School Threatens Children's Well-Being and Undermines Their Education* (Lanham, MD: Rowman and Littlefield, 2015).
20. Molnar and Boninger point out that the Commercialism in Education Research Unit (CERU) reported that in the 2003–2004 school year close to 70 percent of schools turning to commercialism to gain income

earned no money at all and that only four schools earned more than $50,000. In 2012, a report from Public Citizen found that even contracts for advertising that seemed likely to generate big bucks usually brought in only about .03 percent of a school district's budget (Elizabeth Ben-Ishai, "School Commercialism: High Cost, Low Revenues," *Public Citizen*, 2012); see this national survey of the nature and extent of marketing activities in American public schools: Alex Molnar et al., "Marketing of Foods of Minimal Nutritional Value in Schools," *Preventive Medicine* 47, no. 5 (November 2008): 504–7, doi.org/10.1016/j.ypmed.2008.07.019; Brian O. Brent and Stephen Lunden, "Much Ado About Very Little: The Benefits and Costs of School-Based Commercial Activities," *Leadership and Policy in Schools* 8, no. 3 (July 2009): 307–36, doi.org/10.1080/15700760802488619.

21. Ed Winter quoted in Pat Wechsler, "This Lesson Is Brought to You By . . . ," *Business Week*, June 30, 1997, 68.

22. Joel Babbit quoted in Ralph Nader, *Children First: A Parent's Guide to Fighting Corporate Predators* (Washington, DC: Children First, 1996), 64.

23. Consumers Union of the United States, *Captive Kids: A Report on Commercial Pressures on Kids at School* (Yonkers, NY: Consumers Union Education Services, 1995).

24. Helen G. Dixon et al., "The Effects of Television Advertisements for Junk Food Versus Nutritious Food on Children's Food Attitudes and Preferences," *Social Science and Medicine* 65, no. 7 (October 2007): 1311–23.

25. Jeff Cronin and Richard Adcock, "USDA Urged to Protect Kids from Digital Food Marketing on Online Learning Platforms," *Center for Science in the Public Interest*, July 13, 2020.

26. Alex Molnar, "The Commercial Transformation of Public Education," *Journal of Education Policy* 21, no. 5 (September 1, 2006): 632, doi.org/10.1080/02680930600866231. For a good description of how this works, see Rachel Cloues, "The Library That Target Built," *Rethinking Schools* 8, no. 4 (Summer 2014).

27. Molnar, *Journal of Education Policy* 21: 632.

28. Bradley S. Greenberg and Jeffrey E. Brand, "Television News Advertising in Schools: The 'Channel One' Controversy," *Journal of Communication* 43, no. 1 (1993): 143–51.

29. See John Dewey, *Democracy and Education* (New York: Macmillan, 1916).

30. Bernays is quoted in Alex Molnar and Faith Boninger, "The Commercial Transformation of America's Schools," *Phi Delta Kapan* (blog),

September 21, 2020, kappanonline.org/commercial-transformation-americas-schools-molnar-boninger.
31. Stuart Ewan, *Captains of Consciousness: Advertising and the Social Roots of Consumer Culture* (New York: McGraw Hill, 1976), 54–55.
32. Molnar and Boninger, "Commercial Transformation of America's Schools."
33. Robert Bryce, "Marketing Wars Enter Schoolyard Contracts Cut by Coke and Nike Stir Ethical Concerns About Commerce on Campus," *Christian Science Monitor*, August 19, 1997, sec. UNITED STATES.
34. "Mission and History: American Coal Council," American Coal Council, www.americancoalcouncil.org/page/mission.
35. Erik W. Robelen, "Scholastic to Scale Back Corporate Sponsorships," *Education Week*, Bethesda 30, no. 37 (August 10, 2011): 5.
36. Tamar Lewin, "Coal Curriculum Called Unfit for 4th Graders," *New York Times*, May 11, 2011.
37. "Stem Classroom Activities and Resources: The Fracking Debate." Shell Company, www.shell.us/sustainability/energize-your-future-with-shell/stem-classroom-activities/_jcr_content/par/textimage_1406346029.stream/1519761023329/39c1976bd9e3fd78bf11d90616dfd18e9545ebe9/eyf-fracking-debate.pdf.
38. "Changing Environments," BP Educational Services, bpes.bp.com/changing-environments-video-and-comprehension.
39. "Voluntarily Funded by the People of Oklahoma Oil and Natural Gas," Oklahoma Energy Resources Board, Funding (OERB), www.oerb.com/about/funding.
40. According to the 2019 index of polluters from the Political Economy Research Institute at the University of Massachusetts, Koch industries ranks seventeenth in air pollution, twenty-second in greenhouse gas emissions, and thirteenth in water pollution; Matthew Baylor, "Combined Toxic 100 Greenhouse 100 Indexes," Political Economy Research Institute, www.peri.umass.edu/combined-toxic-100-greenhouse-100-indexes-current.
41. Katie Worth, "Climate Change Skeptic Group Seeks to Influence 200,000 Teachers," *Frontline*, March 28, 2017, pbs.org/wgbh/frontline/article/climate-change-skeptic-group-seeks-to-influence-200000-teachers.
42. "Frequently Asked Questions," Bill of Rights Institute, billofrightsinstitute.org/about-bri/faq.
43. "Educator Hub," Bill of Rights Institute, billofrightsinstitute.org/educate.
44. Roberto A. Ferdman, "McDonald's Quietly Ended Controversial

Program That Was Making Parents and Teachers Uncomfortable," *Washington Post*, May 13, 2016.
45. Discovery Education, "Pathway Financial in Schools," www.pathwayinschools.com.
46. Discovery Education, "Pathway to Financial Success in Schools."
47. "What Happens If My Credit Card Payment Is Late?," Discover, Credit Resource Center, www.discover.com/credit-cards/resources/so-my-credit-card-is-past-due.
48. Here are just a few examples: for educators, see Shirley Rush, "Children's Needs Versus Wants," NC Cooperative Extension, Hoke County Center, hoke.ces.ncsu.edu/2016/09/childrens-needs-versus-wants/; for early childhood experts, see Cheryl Flanders, "'But I WAAAANT It!' How to Teach Kids the Difference Between Wants and Needs," *Kinder Care* (blog), www.kindercare.com/content-hub/articles/2017/december/teach-difference-wants-and-needs; Tom Popomaronis, "Warren Buffett: This is the No. 1 Mistake Parents Make When Teaching Kids About Money," CNBC, Make It, July 30, 2019, last modified July 9, 2020, www.cnbc.com/2019/07/30/warren-buffett-this-is-the-no-1-mistake-parents-make-when-teaching-kids-about-money.html.
49. Practical Money Skills, "Lesson Two: Spending Plans," www.practicalmoneyskills.com/assets/pdfs/lessons/lev_1/2_complete.pdf.
50. Practical Money Skills, "Lesson Two: Spending Plans."
51. Jim Higlym, *Men's Health* 28, no. 10 (December 2013): 88.
52. Jonny Tiernan, "Impossible Foods to Reach New Generation Through US School Nutrition Programs," *CleanTechnica*, May 17, 2021. cleantechnica.com/2021/05/17/impossible-foods-to-reach-new-generation-through-us-school-nutrition-programs.
53. Anna Lappé, "Impossible Foods, Impossible Claims, Real Food Media," *Medium*, July 22, 2019, medium.com/real-food-media/impossible-foods-impossible-claims-c10ef2e457ed.
54. Center for Food Safety, "Center for Food Safety Filed Lawsuit Against FDA Challenging Decision to Approve Genetically Engineered Soy Protein Found in the Impossible Burger," new release, March 18, 2020, centerforfoodsafety.org/press-releases/5961/center-for-food-safety-filed-lawsuit-against-fda-challenging-decision-to-approve-genetically-engineered-soy-protein-found-in-the-impossible-burger.
55. Pavni Diwanji, "'Be Internet Awesome': Helping Kids Make Smart Decisions Online," *Google* (blog), June 6, 2017, www.blog.google/technology/families/be-internet-awesome-helping-kids-make-smart-decisions-online.

56. "Child and Consumer Advocates Urge FTC to Investigate and Bring Action Against Google for Excessive and Deceptive Advertising Directed at Children: So-Called 'Family-Friendly' YouTube Kids App Combines Commercials and Videos, Violating Long-Standing Safeguards for Protecting Children," Campaign for Commercial-Free Childhood, April 6, 2015, commercialfreechildhood.org/advocates-file-ftc-complaint-against-googles-youtube-kids.
57. Alistair Barr, "Google's YouTube Kids App Criticized for 'Inappropriate Content,'" *Wall Street Journal*, May 19, 2015, blogs.wsj.com/digits/2015/05/19/googles-youtube-kids-app-criticized-for-inappropriate-content.
58. Benjamin Herold, "Big Data or Big Brother?," *Education Week* 35, no. 17 (January 13, 2016), search.proquest.com.ezp-prod1.hul.harvard.edu/docview/1759529343?accountid=11311.
59. Richard Serra et al., "You're Not the Customer; You're the Product," Quote Investigator, July 16, 2017, quoteinvestigator.com/2017/07/16/product.
60. Jim Seale and Nicole Schoenberger, "Be Internet Awesome: A Critical Analysis of Google's Child-Focused Internet Safety Program," *Emerging Library & Information Perspectives* 1 (2018), 34–58, doi.org/10.5206/elip.v1i1.366.
61. drive.google.com/drive/folders/15QvJEkzq0q7dFVIYABo6BrfKB2HRyVSi.
62. *ISTE 2020 Annual Report*, International Society for Technology in Education (ISTE), 18, iste.org/about/iste-story/annual-report.
63. "Hosted Activities," International Society for Technology in Education (ISTE), ISTE Live 20, conference.iste.org/2020/program/hosted_activities.php; "ISTE 2020 Conference and Expo Sponsorship Opportunities," ISTE, February 8, 2020, conference.iste.org/2020/exhibitors/pdfs/ISTE20SponsorshipOpportunities.pdf.
64. "Interested in Becoming a Member?," Family Online Safety Institute (FOSI), www.fosi.org/membership#member-form; "2016 Annual Conference," FOSI, www.fosi.org/events/2016-annual-conference; "2017 Annual Conference," FOSI, www.fosi.org/events/2017-annual-conference; "2018 Annual Conference," FOSI, www.fosi.org/events/2018-annual-conference; "2019 Annual Conference," FOSI, www.fosi.org/events/2019-annual-conference.
65. "Sponsors and Partners," National PTA, www.pta.org/home/About-National-Parent-Teacher-Association/Sponsors-Partners.
66. For an excellent discussion of how and why what children learn in

schools is never neutral and how that relates to commercialism, see Trevor Norris, *Consuming Schools and the End of Politics* (Toronto: University of Toronto Press, 2011), 48–60.

67. Howard Zinn, *The Politics of History: With a New Introduction* (Bloomington: University of Illinois Press, 1990), 24.

CHAPTER 10: BIG TECH GOES TO SCHOOL

Lisa Cline, email message to author, July 9, 2021.

1. *Let's Make a Sandwich*, produced by Simmel-Meservey in collaboration with the American Gas Association (1950), www.youtube.com/watch?v=Xz4SYkwSxTM.
2. "The Story of Lubricating Oil: Standard Oil Educational Film," Periscope Films, 1949, www.youtube.com/watch?v=4noZ0OaSyFc.
3. "Wikipedia: Pollution of the Hudson River," Wikimedia Foundation, last modified March 27, 2021, 23:50, en.wikipedia.org/wiki/Pollution_of_the_Hudson_River.
4. Faith Boninger, "The Demise of Channel One," interview by *NEPC Newsletter* (Boulder, CO: National Education Policy Center), August 2, 2018, nepc.colorado.edu/publication/newsletter-channel-one-080218.
5. "About Channel One News," Channel One, www.channelone.com/common/about.
6. William Hoynes, "News for a Captive Audience: An Analysis of Channel One," *Extra!* (New York: FAIR, May/June 1997): 11–17; Nancy Nelson Knupfer and Peter Hayes, "The Effects of Channel One Broadcast on Students' Knowledge of Current Events," in *Watching Channel One*, ed. Ann DeVaney (Albany, NY: SUNY Press, 1994), 42–60; Mark Miller, "How to Be Stupid: The Teachings of Channel One" (Paper prepared for Fairness and Accuracy in Reporting [FAIR], January 1997), 1, files.eric.ed.gov/fulltext/ED405627.pdf.
7. Roy F. Fox, *Harvesting Minds: How TV Commercials Control Kids* (Westport, CT: Praeger, 1996), 92.
8. Bradley S. Greenberg and Jeffrey E. Brand, "Channel One: But What About the Advertising?," *Educational Leadership* 51, no. 4 (December 1993): 56–58; Bradley S. Greenberg and Jeffrey E. Brand, "Commercials in the Classroom: The Impact of Channel One Advertising," *Journal of Advertising Research* 34, no. 1 (1994): 18–27.
9. Faith Boninger, Alex Molnar, and Michael Barbour, *Issues to Consider Before Adopting a Digital Platform or Learning Program* (Boulder,

CO: National Education Policy Center, September 24, 2020), section 3; Roxana Marachi and Lawrence Quill, "The Case of Canvas: Longitudinal Datafication Through Learning Management Systems," *Teaching in Higher Education* 25, no. 4 (April 29, 2020): 418–34; Andrea Peterson, "Google Is Tracking Students as It Sells More Products to Schools, Privacy Advocates Warn," *Washington Post*, December 28, 2015.

10. Criscillia Benford, "Should Schools Rely on EdTech?," *Fair Observer*, August 12, 2020, www.fairobserver.com/region/north_america/criscillia-benford-ed-tech-educational-technology-education-news-tech-coronavirus-covid-19-lockdown-world-news-68173.

11. Organization for Economic Cooperation and Development, "Students, Computers and Learning: Making the Connection," *PISA* (OECD Publishing, September 14, 2015), 3, doi.org/10.1787/19963777.

12. Jake Bryant et al., "New Global Data Reveal Education Technology's Impact on Learning," McKinsey and Company, June 12, 2020, www.mckinsey.com/industries/public-and-social-sector/our-insights/new-global-data-reveal-education-technologys-impact-on-learning.

13. This comes from two sources: analysis by REBOOT of data collected by the National Assessment of Educational Progress in 2017, "Does Educational Technology Help Students Learn?," June 6, 2019, reboot-foundation.org/does-educational-technology-help-students-learn; Helen Lee Bouygues, "Addendum: The 2019 NAEP Data on Technology and Achievement Outcomes," Reboot Foundation, November 22, 2019, reboot-foundation.org/wp-content/uploads/_docs/2019_NAEP_Data_Update_Memo.pdf.

14. Natasha Singer, "How Google Took Over the Classroom," *New York Times*, May 13, 2017.

15. Singer, "How Google Took Over the Classroom."

16. Matthew Lynch, "Top 9 Must Have Personalized Learning Apps, Tools, and Resources," *Tech Edvocate*, August 7, 2017, www.thetechedvocate.org/top-9-must-personalized-learning-apps-tools-resources.

17. See for example Alex Molnar's and Faith Boninger's 2020 critique of the immensely popular Summit Learning Platform, included in a *Washington Post* article by Valerie Strass. What's particularly interesting about this piece is that Strauss includes their concerns, Summit's response, as well as Molnar's and Boninger's rebuttal of that response. Valerie Strauss, "New Concerns Raised About a Well-Known Digital Learning Platform," *Washington Post*, June 26, 2020.

18. Sam Garin, "A Statement on EdTech and Education Policy During the

Pandemic," Campaign for a Commercial-Free Childhood, August 11, 2020, commercialfreechildhood.org/edtech_statement; "the value of quality teacher-driven instruction," see e.g., Saro Mohammed, "Tech or No Tech, Effective Learning Is All About Teaching," *Brookings*, September 6, 2018, www.brookings.edu/blog/brown-center-chalkboard/2018/09/06/tech-or-no-tech-effective-learning-is-all-about-teaching/; "no credible research supporting industry claims," see Alex Molnar et al., "Virtual Schools in the U.S. 2019," National Education Policy Center, May 28, 2019, nepc.colorado.edu/publication/virtual-schools-annual-2019; Organization for Economic Cooperation and Development, *Students, Computers and Learning* (Paris: OECD Publishing, 2015), doi.org/10.1787/19963777.

19. Prodigy Education, "Prodigy Education (Formerly Prodigy Game) Recognized as One of Canada's Top 3 Fastest-Growing Companies," *Cision PR Newswire*, September 16, 2019, www.prnewswire.com/news-releases/prodigy-education-formerly-prodigy-game-recognized-as-one-of-canadas-top-3-fastest-growing-companies-300918004.html.
20. Prodigy Education, "Fastest-Growing Companies."
21. Prodigy Education, "Fastest-Growing Companies."
22. Prodigy Education, "Prodigy Education Expands Its Market Leadership in Game-Based Learning to English Language Arts," *Cision PR Newswire*, October 26, 2021. www.prnewswire.com/news-releases/prodigy-education-expands-its-market-leadership-in-game-based-learning-to-english-language-arts-301408101.html; Prodigy Education, "Prodigy English," www.prodigygame.com/main-en/prodigy-english.
23. Rheta Rubenstein, email message to author, September 27, 2020.
24. Prodigy, "Play the Game," www.prodigygame.com/main-en.
25. Campaign for a Commercial-Free Childhood et al., "Request for Investigation of Deceptive and Unfair Practices by the Edtech Platform Prodigy" (letter of complaint to the Federal Trade Commission, February 19, 2021), 18, fairplayforkids.org/wp-content/uploads/2021/02/Prodigy_Complaint_Feb21.pdf.
26. Prodigy, "Premium Memberships: Inspire Your Child to Love Learning. Support Them Along the Way," Membership, www.prodigygame.com/Membership.
27. Prodigy, "Premium Memberships: Inspire Your Child to Love Learning. Support Them Along the Way," Membership, www.prodigygame.com/Membership.

28. Prodigy, "Give Your Mighty Learner an Epic Gift with a Premium Membership/The Premium Membership Works," www.prodigygame.com/in-en/prodigy-memberships
29. "Prodigy Meets ESSA Tier 3 in a Study by Johns Hopkins University," Prodigy, www.prodigygame.com/main-en/research.
30. The Center for Research and Reform in Education (CRRE) at the Johns Hopkins University, "Evaluation of Prodigy: Key Findings," Prodigy (2020), 3, prodigy-website.cdn.prismic.io/prodigy-website/9ee58b31-3537-42c3-9eff-f8e6b16db030_Prodigy-Evaluation-Key-Findings_Final.pdf.
31. "Parents Accounts," Prodigy, www.prodigygame.com/pages/parents.
32. Rebecca, French teacher, product review for Prodigy, October 3, 2015, www.edsurge.com/product-reviews/prodigy/educator-reviews.
33. Campaign for a Commercial-Free Childhood et al., "Request for Investigation of Deceptive and Unfair Practices by the Edtech Platform Prodigy."
34. Conversation with author, March 10, 2021.
35. Leilani Cauthen, "Just How Large Is the K12 Ed-Tech Market," *EdNews Daily*, www.ednewsdaily.com/just-how-large-is-the-k12-ed-tech-market.
36. Valerie J. Calderon and Margaret Carlson, "Educators Agree on the Value of EdTech," *Gallup*, Education, September 12, 2019, www.gallup.com/education/266564/educators-agree-value-tech.aspx.
37. Calderon and Carlson, "Educators Agree on the Value of EdTech."
38. Calderon and Carlson, "Educators Agree on the Value of EdTech."
39. Natalie Wexler, "How Classroom Technology Is Holding Students Back," *MIT Technology Review*, December 19, 2019, www.technologyreview.com/s/614893/classroom-technology-holding-students-back-edtech-kids-education/?eType=EmailBlastContent&eId=d824edf8-1903-4acd-a474-59b0bd250982.
40. Jacquelyn Ottman, "The Four E's Make Going Green Your Competitive Edge," *Marketing News* 26, no. 3 (February 3, 1992): 7.
41. Bruce Watson, "The Troubling Evolution of Corporate Greenwashing," *The Guardian*, August 20, 2016.
42. See John F. Pane, "Strategies for Implementing Personalized Learning While Evidence and Resources Are Underdeveloped," *Perspective* (Santa Monica, CA: Rand Corporation, October 2018), 7, doi.org/10.7249/PE314.
43. Alfie Kohn, "Four Reasons to Worry About 'Personalized Learning,'" *Psychology Today*, February 24, 2015, www.psychologytoday.com/us

/blog/the-homework-myth/201502/four-reasons-worry-about-personalized-learning.
44. See Howard Gardner, *The Unschooled Mind: How Children Learn and Schools Should Teach* (New York: Basic Books, 1991).
45. Faith Boninger, Alex Molnar, and Christopher Saldaña, "Personalized Learning and the Digital Privatization of Curriculum and Teaching," National Education Policy Center Boulder, April 30, 2019, nepc.colorado.edu/publication/personalized-learning.
46. "Game-Based Learning Market Worth $29.7 Billion by 2026—Exclusive Report by MarketsandMarketsTM," *PR Newswire*, www.prnewswire.com/news-releases/game-based-learning-market-worth-29-7-billion-by-2026--exclusive-report-by-marketsandmarkets-301462512.html.
47. Ben Feller, "Video Games Can Reshape Education," *NBC News*, October 17, 2006; Henry Kelly, *Harnessing the Power of Games for Learning* (Presummit paper for the Summit on Educational Games, Federation of American Scientists, 2005), www.informalscience.org/sites/default/files/Summit_on_Educational_Games.pdf
48. My colleague Diane Levin coined the term *problem-solving deficit disorder* as a condition of a modern childhood where children don't get enough time for creative play. See Barbara Meltz, "There Are Benefits to Boredom," *Boston Globe*, January 22, 2004; see Lev S. Vygotsky, "Play and Its Role in the Mental Development of the Children," in *Play: Its Role in Development and Evolution*, ed. Jerome S. Bruner, Alison Jolly, and Kathy Sylva (New York: Basic Books, 1976), 536–52; Kathleen Roskos and Susan B. Neuman, "Play as an Opportunity for Literacy," in *Multiple Perspectives on Play in Early Childhood*, ed. Olivia N. Saracho and Bernard Spodek (Albany, NY: SUNY Press, 1998), 100–16; I wrote extensively about the benefits of play in *The Case for Make Believe: Saving Play in a Commercialized World* (New York: New Press, 2009). Also see the work of Diane Levin and Nancy Carlsson-Paige, *The War Play Dilemma*, 2nd ed. (New York: Teachers College Press, 2006) and *Who's Calling the Shots* (St. Paul, MN: New Society, 1987).
49. "Prodigy Education | Make Learning Math Fun!," Prodigy Education, 2020, www.youtube.com/watch?v=O1_V75nK67M.
50. "Gamification: Learning Made Fun with Genially," The Techie Teacher, April 20, 2020, www.thetechieteacher.net/2020/04/gamification-learning-made-fun-with.html.
51. See Rishi Desai, MD, Khan Academy Fellow at "Khan Academy

Gamification: Making Learning Fun," YouTube video, September 10, 2013, www.youtube.com/watch?v=2EZcpZSy58o.
52. "Fun, n. and adj.," in *OED Online* (Oxford University Press), www.oed.com/view/Entry/75467.
53. Kathy Hirsh-Pasek et al., "Putting Education in 'Educational' Apps: Lessons from the Science of Learning," *Psychological Science in the Public Interest* 16, no. 1 (2015): 3–34.
54. Kathy Hirsh-Pasek et al., "A Whole New World: Education Meets the Metaverse," Brooking Institute, February, 2022, www.brookings.edu/research/a-whole-new-world-education-meets-the-metaverse.
55. See Faith Boninger, Alex Molnar, and Michael K. Barbour, "Issues to Consider Before Adopting a Digital Platform or Learning Program," National Education Policy Center, September 24, 2020, nepc.colorado.edu/publication/virtual-learning. Their brief raises important questions and concerns worth considering about edtech.

Chapter 11: Is That Hope?

Alex Molnar and Faith Boninger, "The Commercial Transformation of America's Schools," *Phi Delta Kappan*, September 21, 2020, kappanonline.org/commercial-transformation-americas-schools-molnar-boninger.

1. Russell Banks, "Feeding Moloch: The Sacrifice of Children on the Altar of Capitalism," Ingersoll Lecture, Harvard Divinity School, November 5, 2014, www.youtube.com/watch?v=Kv3Rbq5WIK4.
2. Nipa Saha, "Advertising Food to Australian Children: Has Self-Regulation Worked?," *Journal of Historical Research in Marketing* 12, no. 4 (October 19, 2020): 525–50; Nancy Fink Huehnergarth, "Coca-Cola Skirts Its Own Pledge Not to Market to Young Kids, Says Report," *Forbes*, www.forbes.com/sites/nancyhuehnergarth/2016/05/20/coca-cola-skirts-its-own-pledge-not-to-market-to-young-kids-says-report.
3. Matthew Chapman, "Big Tobacco Targets Youth with $1.4 Billion Marketing Campaign," *Children's Health Defense* (blog), March 15, 2021, childrenshealthdefense.org/defender/big-tobacco-targets-youth.
4. Children's Food and Beverage Initiative, Better Business Bureau, bbbprograms.org/programs/all-programs/cfbai#.
5. Jennifer A. Emond et al., "Unhealthy Food Marketing on Commercial Educational Websites: Remote Learning and Gaps in Regulation," *American Journal of Preventive Medicine* 60, no. 4 (2021): 587–91.
6. B. Shaw Drake and Megan Corrarino, "U.S. Stands Alone: Not Signing

U.N. Child Rights Treaty Leaves Migrant Children Vulnerable," *Huff-Post*, October 13, 2015, www.huffpost.com/entry/children-migrants-rights_b_8271874.
7. David Cribb, "Pussycat Dolls Cancelled After Complaints," *Digital Spy*, May 25, 2006; United Press, "Hasbro Backs off Pussycat Dolls," May 25, 2006.
8. Tamar Lewin, "No Einstein in Your Crib? Get a Refund," *New York Times*, October 24, 2009.
9. Scott van Voorhis, "BusRadio Gains Industry Ratings, Criticism for Pushing Ads on Kids," *Boston Herald*, July 8, 2008.
10. "McDonald's Gets F Grade in Florida," Corpwatch, www.corpwatch.org/article/mcdonalds-gets-f-grade-florida.
11. "You Did It: Facebook Pauses Plans for Instagram for Kids," Fairplay, fairplayforkids.org/you-did-it-facebook-pauses-plans-for-instagram-for-kids.
12. John Davidson, "Finally, the World Takes on Big Tech," *Australian Financial Review*, December 23, 2020, www.proquest.com/docview/2471684612/citation/F64199E3B2C84FDCPQ/1.
13. Ari Ezra Waldman, "How Big Tech Turns Privacy Laws into Privacy Theater," *Slate*, December 2, 2021.
14. "Age Appropriate Design: A Code of Practices for Online Services," Information Commissioner's Office, ico.org.uk/for-organisations/guide-to-data-protection/ico-codes-of-practice/age-appropriate-design-a-code-of-practice-for-online-services.
15. "Age Appropriate Design."
16. Maite Fernández Simon, "Australia Proposes Parental Consent for Children Under 16 on Social Media," *Washington Post*, October 26, 2021; Swedish Data Protection Authority, "The New Directive Will Reinforce the Child and Youth Rights," *Privacy 365*, October 23, 2020, www.privacy365.eu/en/by-the-swedish-data-protection-authority-the-new-directive-will-reinforce-the-child-and-youth-rights.
17. Angela Campbell, "Protecting Kids Online: Internet Privacy and Manipulative Marketing," *Testimony Before the Senate Committee on Commerce, Science, and Transportation's Subcommittee on Consumer Protection, Product Safety, and Data Security*, May 18, 2021, www.commerce.senate.gov/services/files/9935A07E-AC61-4CFD-A422-865D89C54EA3.
18. Campbell, "Protecting Kids Online."
19. "Full Transcript of Biden's State of the Union Address," *New York Times*, March 2, 2022.

20. Protecting the Information of Our Vulnerable Children and Youth Act (Kids PRIVCY Act), H.R. 4801, 117 Cong., 1st Sess.§__ (2021). www.congress.gov/117/bills/hr4801/BILLS-117hr4801ih.pdf. For a summary of the bill's salient points, see castor.house.gov/news/documentsingle.aspx?DocumentID=403677.
21. "Rep. Castor Reintroduces Landmark Kids PRIVCY Act to Strengthen COPPA, Keep Children Safe Online," Representative Kathy Castor, July 29, 2021, castor.house.gov/news/documentsingle.aspx?DocumentID=403677.
22. Shannon Bond, "New FTC Chair Lina Khan Wants to Redefine Monopoly Power for the Age of Big Tech," NPR, July 1, 2021, www.npr.org/2021/07/01/1011907383/new-ftc-chair-lina-khan-wants-to-redefine-monopoly-power-for-the-age-of-big-tech.
23. "Blackburn & Blumenthal Introduce Comprehensive Kids' Online Safety Legislation," February 16, 2022, www.blackburn.senate.gov/2022/2/blackburn-blumenthal-introduce-comprehensive-kids-online-safety-legislation.
24. Edward Markey, "Text-S.2918-117th Congress (2021–2022): KIDS Act," Congress.gov, September 30, 2021, www.congress.gov/bill/117th-congress/senate-bill/2918/text. The bill is called the "Kids Internet Design and Safety Act."
25. A summary of the KIDS Act can be found at: "Senators Markey and Blumenthal, Rep. Castor Reintroduce Legislation to Protect Children and Teens from Online Manipulation and Harm, U.S. Senator Ed Markey of Massachusetts," www.markey.senate.gov/news/press-releases/senators-markey-and-blumenthal-rep-castor-reintroduce-legislation-to-protect-children-and-teens-from-online-manipulation-and-harm.
26. "Senators Markey and Blumenthal, Rep. Castor Reintroduce Legislation to Protect Children and Teens from Online Manipulation and Harm, U.S. Senator Ed Markey of Massachusetts," www.markey.senate.gov/news/press-releases/senators-markey-and-blumenthal-rep-castor-reintroduce-legislation-to-protect-children-and-teens-from-online-manipulation-and-harm.
27. "State Policies to Prevent Obesity," *State of Childhood Obesity* (blog), American Academy of Pediatrics, stateofchildhoodobesity.org/state-policy/policies/screentime. Day.
28. www.revisor.mn.gov/bills/bill.php?b=senate&f=SF237&ssn=0F&y=2021.
29. "Your State Legislation | Media Literacy Now," January 24, 2014, medialiteracynow.org/your-state-legislation-2.

CHAPTER 12: RESISTANCE PARENTING

Rachel Franz, "Safe, Secure, and Smart—Preschool Apps," *Fairplay* (blog), fairplayforkids.org/pf/safe-secure-smart-apps.

1. Notable recommendations include those of the World Health Organization (www.who.int/publications/i/item/9789241550536), the American Academy of Pediatrics (doi.org/10.1542/peds.2016-2591; doi.org/10.1542/peds.2016-2592), Australian Government Department of Health (www.health.gov.au/health-topics/physical-activity-and-exercise/physical-activity-and-exercise-guidelines-for-all-australians/for-infants-toddlers-and-preschoolers-birth-to-5-years; www.health.gov.au/health-topics/physical-activity-and-exercise/physical-activity-and-exercise-guidelines-for-all-australians/for-children-and-young-people-5-to-17-years), Canadian Paediatric Society (cps.ca/en/documents//position//screen-time-and-young-children), Indian Association of Pediatrics (www.indianpediatrics.net/sep2019/sep-773-788.htm), and Italian Pediatric Society (doi.org/10.1186/s13052-018-0508-7; doi.org/10.1186/s13052-019-0725-8).
2. Russell Viner et al., "The Health Impacts of Screen Time: A Guide for Clinicians and Parents," *The Royal College of Paediatrics and Child Health*, January 2019, www.rcpch.ac.uk/sites/default/files/2018-12/rcpch_screen_time_guide_-_final.pdf.
3. Brae Ann McArthur et al., "Longitudinal Associations Between Screen Use and Reading in Preschool-Aged Children, *Pediatrics*, 147 no. 6 (June 2021), publications.aap.org/pediatrics/article-abstract/147/6/e2020011429/180273/Longitudinal-Associations-Between-Screen-Use-and?redirectedFrom=fulltext; Dimitri Christakis and Fred Zimmerman, "Early Television Viewing Is Associated with Protesting Turning off the Television at Age 6," *Medscape General Medicine* 8, no. 2 (2006): 63, www.medscape.com/viewarticle/531503.
4. Tiffany G. Munzer et al., "Differences in Parent-Toddler Interactions with Electronic Versus Print Books," *Pediatrics* 143, no. 4 (April 2019): e20182012, doi.org/10.1542/peds.2018-2012.
5. Marisa Meyer et al., "How Educational Are 'Educational' Apps for Young Children? App Store Content Analysis Using the Four Pillars of Learning Framework," *Journal of Children and Media* (February 2021), doi.org/10.1080/17482798.2021.1882516.
6. Kathy Hirsh-Pasek et al., "Putting Education in 'Educational' Apps: Lessons from the Science of Learning," *Psychological Science in the Public Interest* 16, no. 1 (2015): 3–34, doi.org/10.1177/1529100615569721.

7. Zoe Forsey, "Meet Bill and Melinda Gates's Children—But They Won't Inherit Their Fortune," *Mirror*, May 4, 2021, www.mirror.co.uk/news/us-news/meet-bill-melinda-gatess-impressive-24035569.
8. Deborah Richards, Patrina H.Y. Caldwell, and Henry Go, "Impact of Social Media on the Health of Children and Young People," *Journal of Paediatrics and Child Health* 51, no. 12 (2015): 1152–57, doi.org/10.1111/jpc.13023.
9. danah boyd, "Panicked About Kids' Addiction to Tech?," *NewCo Shift* (blog), January 25, 2018, medium.com/newco/panicked-about-kids-addiction-to-tech-88b2c856bf1c.
10. Emily Cherkin, "Babies and Screentime," Screentime Consultant, May 11, 2020, www.thescreentimeconsultant.com/blog/babies-and-screentime.
11. Josh Golin, email message to author, September 10, 2021.
12. Lauren Hale and Stanford Guan, "Screen Time and Sleep Among School-Aged Children and Adolescents: A Systematic Literature Review," *Sleep Medicine Reviews* 21 (2015): 50–58, doi.org/10.1016/j.smrv.2014.07.007.

CHAPTER 13: MAKING A DIFFERENCE FOR EVERYBODY'S KIDS

Joseph Bates, email to author, February 2, 2022.

1. According to the nonprofit Open Secrets, in the first six months of 2021, Facebook spent $9.56 million on lobbying, National Amusements, which owns Viacom, spent $2.43 million, Disney spent $2.38 million, and Alphabet, which owns Google, spent $5.66 million. See Open Secrets, July 23, 2021, www.opensecrets.org/federal-lobbying/summary.
2. Adam Hochschild, *Bury the Chains: Prophets and Rebels in the Fight to Free an Empire's Slaves* (Boston: Houghton Mifflin, 2005).
3. "Women's Suffrage in the United States," *Wikipedia*, December 8, 2021, en.wikipedia.org/w/index.php?title=Women%27s_suffrage_in_the_United_States&oldid=1059301611.
4. "Same-Sex Marriage in the United States," *Wikipedia*, October 24, 2021, en.wikipedia.org/w/index.php?title=Same-sex_marriage_in_the_United_States&oldid=1051634688.
5. Vincent Harding, *There Is a River: The Black Struggle for Freedom* (New York: Harcourt Brace Jovanovich, 1981).
6. Ariel Schwartz, "Computer Algorithms Are Now Deciding Whether

Prisoners Get Parole," *Business Insider*, December 15, 2015, www.businessinsider.com/computer-algorithms-are-deciding-whether-prisoners-get-parole-2015-12.
7. Noel Maalouf et al., "Robotics in Nursing: A Scoping Review," *Journal of Nursing Scholarship* 50, no. 6 (2018): 590–600, doi.org/10.1111/jnu.12424.
8. Natasha Singer, "Deciding Who Sees Students' Data," *New York Times*, October 5, 2013.
9. Natasha Singer, "InBloom Student Data Repository to Close," *Bits Blog* (blog), April 21, 2014, https://bits.blogs.nytimes.com/2014/04/21/inbloom-student-data-repository-to-close.
10. Elaine Meyer, "How Gen Z Is Fighting Back Against Big Tech," *Yes! Media*, November 24, 2021.
11. Log Off: A Movement Dedicated to Rethinking Social Media by Teens for Teens, www.logoffmovement.org.
12. "Sponsors and Partners—About PTA, National PTA," www.pta.org/home/About-National-Parent-Teacher-Association/Sponsors-Partners.
13. Fairplay, "Faith Leaders Letter to Mark Zuckerberg," fairplayforkids.org/wp-content/uploads/2022/02/FaithIG.pdf, January 8, 2022. The quote in the following paragraph is from this same letter.
14. Fairplay does a great job of articulating the differences between what they call a "branded childhood" and a "childhood beyond brands." See "What Is a Childhood Beyond Brands?," *Fairplay* (blog), fairplayforkids.org/beyond-brands.

Afterword

1. Adam Satariano, "E.U. Takes Aim at Social Media's Harms with Landmark New Law," *New York Times*, April 22, 2022.
2. Cecilia Kang, "As Europe Approves New Tech Laws, the U.S. Falls Further Behind," *New York Times*, April 22, 2022.
3. "LEGO Teams Up with Cary's Epic Games to Help Build a Kid-Safe Virtual World | WRAL TechWire," April 7, 2022, wraltechwire.com/2022/04/07/lego-teams-up-with-carys-epic-games-to-help-build-a-kid-safe-virtual-world.
4. Fairplay, "Designing_for_disorder.Pdf," April, 2022, fairplayforkids.org/wp-content/uploads/2022/04/designing_for_disorder.pdf.
5. Emma Goldberg, "'Techlash' Hits College Campuses," *New York Times*, January 11, 2020.

6. Colleen Mcclain, "How Parents' Views of Their Kids' Screen Time, Social Media Use Changed During COVID-19," *Pew Research Center* (blog), www.pewresearch.org/fact-tank/2022/04/28/how-parents-views-of-their-kids-screen-time-social-media-use-changed-during-covid-19.

Index

ABI Technologies, 131
Accountable Tech, 254
ACM Queue, 146
"active" and "passive" media, 216
Acton-Boxborough Regional School
    District's edtech policy, 232–33,
    245–48
  digital tools criteria checklist, 248
  letter to teachers, 245–47
Administrative Theory and Praxis
    (journal), 69–70
Advertising Age, 69
Age-Appropriate Design Code (Great
    Britain), 200–201, 203
Al Otro Lado's Border Rights Project,
    153–54
Aladdin (film), 159
Alana Institute, 254
Algorithmic Justice League, 254
algorithms, xv, 45, 48–49, 50, 137–41,
    147–53
  definitions, 45
  Google's search engines, 146–51
  machine-learning, 45
  personalized recommendations, 45,
    48–49, 50, 137–41
  predictive, 45, 48–49, 50, 137–41
  racist biases and stereotypes, xv, 146–53
  social media sites and, 151–53

Algorithms of Oppression (Nobel), 149
AlgoTransparency, 48–49
Alliance for Childhood, 25n
allowances, children's, 226
Almon, Joan, 13, 25n
Alphabet, 102
Alter, Adam, 33
Amazon, 59, 77, 133–41, 165
  Amazon Kid+, 136
  devices marketed to children, 134–36,
    151
  digital assistants, 133–34, 137–41, 151
  Echo Dot Kids Edition, 134–41, 151
  Kindle e-reader, 134
  one-click buying, 77
  persuasive design, 46
  predictive algorithms and personalized
    recommendations, 137–41
Amazon Alexa (voice-activated digital
    assistant), 133–34, 137–41, 151
  "Alexa, I'm bored" feature, 139–40
  and Echo Dot Kids Edition, 134–41,
    151
  predictive algorithms and personalized
    recommendations, 137–41
  search feature and racism, 151
  "skills," 134, 141
American Academy of Pediatrics (AAP),
    43, 52–53

INDEX

American Coal Foundation, 170
American Gas Association, 182
American Girl dolls, 139
American Psychological Association (APA), 51–52
American Speech-Language-Hearing Association, 41, 238
Amnesty International, 68
Apple
  app store, 216
  branding and brand loyalty, 67–68, 165
  customers' privacy, 68
  digital assistant, 134
  hardware dumping in schools, 184
  iCloud, 68
  in-school advertising (teaching materials), 165
  Super Bowl commercial (1984), 67–68
  See also iPads
apps
  age-appropriate, 216
  commercial-free, 217
  edtech and Prodigy, 181, 184–89, 232–33
  facilitating meaningful learning, 194–95
  "free" apps ("freemiums"), 46, 119, 181, 186–87, 217
  Lego, 4, 93–94, 95, 100, 119
  Lego City, 93–94, 95, 100
  nagware (pop-up ads), 46, 186–87, 189
Aristotle (Mattel smart device), 36–38
artificial intelligence (AI), 66, 135, 141–43
  communicative robots (or "sociable bots"), 141–43, 231
  and parasocial relationships, 142–43
Ascano, Armida, 61–62, 64–65
The Atlantic, 85
attachment, 36–38, 79
  and smart devices, 36–38, 141–43
Australia, 74–75, 93, 94, 118
Australian Financial Review, 200
autoplay settings, 77–78, 128, 204
awe and wonder, 21–24
Axe, 64–65

Baby Einstein videos, 4, 199
Backyard Basecamp, 255
Balloon Pop (app), 119
Banks, Russell, 197
Barbie, 59, 85, 138, 139, 155–57
  Barbie You Can Be Anything, 139
  Black Barbie, 157
  Hello Barbie, 142
  Hispanic Barbie, 155–57, 162
Barton, Christopher, 158
Barton, Joe, 38
Bates, Joseph, 229
Be Internet Awesome (Google SEMS), 176–79
Be Tech Wise with Baby! (American Speech-Language-Hearing Association), 238
Beanie Babies, 84
BEGIN (early learning tech company), 136
behavioral advertising, 46, 136, 195
Benford, Criscillia, 44n, 46n, 183–84
Benjamin, Ruha, xv
Bequelin, Nicholas, 68
Bernays, Edward, 168–69
Bezos, Jeff, 77
Biden, Joe, 203
Big Food, 198. *See also* food industry
Big Tobacco, 198
Bill and Melinda Gates Foundation, 233, 234
Bill of Rights Institute, 172
Billboard charts, 90
Bisiewicz, Amy, 245n
Black Barbie, 157
Black Lives Matter movement, 150, 152
The Bluest Eye (Morrison), 155
Bollywood films, 157
Boninger, Faith, 197
Bookis, Deborah, 245n
books. *See* e-books; reading books
boredom
  the "Alexa, I'm bored" feature, 139–40
  providing opportunities to generate tech-free solutions to, 224–25
Boss Baby (film), 139
Boston Children's Hospital, 3, 27–28
Bowling Alone (Putnam), 132
boyd, danah, 221
BP (British Petroleum), 165, 171
Brand Strategy (magazine), 47
branding, 5, 13–14, 57–76, 89
  brand-licensed children's products and toys, 13–14, 29–30
  brand-loyalty, 66–68, 134, 164, 165, 172, 184
  churches/religious institutions, 71
  corporate-financed advertising in schools, 164, 165, 172, 184

INDEX   319

differentiating brand impression from
   reputation, 66
and edtech, 71, 184
presidential, 67, 69–71
social media and brand tribes, 63–65
social media and self-branding, 5, 72,
   76, 219
social media influencers, 72
Toys "R" Us marketing, 58–61
young children and, 89–90
Bratz dolls, 59, 85
Brazil, 160–61
Britain
   adoption of Age-Appropriate Design
      Code, 200–201, 203
   Quakers' efforts to end of slavery, 230
   screen time recommendations, 211
   studies of touch screen-use by babies
      and toddlers, 42
British Food Journal, 116
British Medical Journal, 103
Bronfenbrenner, Urie, 65, 91
Broussard, Meredith, 150
Brown, Pat, 175
Buffett, Warren, 174
Bush, George H.W., 69
Bush, George W., 69, 91
BuzzFeed, 74

campaign finance reform, 199
Campaign for a Commercial-Free
   Childhood (CCFC), 3. See also
   Fairplay
Campbell, Angela, 202–3
Capital One, 163
capitalism
   consumer, 81
   corporate, 81
   surveillance, 80–81
*Captive Kids: A Report on Commercial
   Pressures on Kids at School* (1995
   Consumers Union report), 167
Carey, George, 120–23
Carlsson-Paige, Nancy, 209, 210
Carson, Rachel, 23
Cartoon Network, 94
*The Case for Make Believe: Saving Play
   in a Commercialized World* (Linn),
   21–22, 107, 131
cell phones. *See* smartphones
Center for American Progress, 166–67
Center for Digital Democracy, 231, 255

Center for Humane Technology, 129–30,
   255
Channel One News, 182–83
Charles Koch Foundation, 172
charter schools, 7
Chaslot, Guillaume, 48–49
Cherkin, Emily, 221
Chewbacca, 141
Chicago Tribune, 90
Children and Media Australia (CMA),
   94–95, 216, 255
Children and Nature Network, 255
Children and Screens: Institute for Digital
   Media and Child Development,
   256
Children's Advertising Review Unit
   (CARU), 116
Children's AID Program out of Boston
   University Medical Center, 3
children's healthy growth and
   development, 11–31, 209–11
   attachment, 36–38, 79, 141–43
   brains and early experience, 12–13,
      81–82, 210
   capacity to feel awe and wonder,
      21–24
   constructing knowledge and making
      meaning, 191–92, 210
   creative play, 14–17, 25–29, 42, 87, 93,
      100, 210–11
   critical thinking, 21, 69, 168, 180, 240
   "holding environments," 15–16, 19–20
   intrinsic motivation, 8, 97–103, 107–8
   language acquisition, 41–42
   learning values, 95–97
   the need for silence (quiet times and
      spaces), 25
   self-regulation, 21, 42, 193
   six principles of (for introducing tech),
      209, 210–11
   transitional objects, 79–80
Children's Online Privacy Protection Act
   (COPPA), 135, 201–3
China, 68
Chromebooks, 35, 183–84
Chua, Amy, 61–62
churches, 71, 238–39
Citizens for Responsibility and Ethics in
   Washington (CREW), 70
civil rights movement, 230
Class Size Matters, 233–34
Clean Waters (1945 GE film), 182

climate change, 83–88
  collectibles and plastics pollution, 83–88
  and corporate-sponsored teaching materials in schools, 170–72, 175–76, 179–80
  and greenwashing, 190
  and Koch-sponsored science materials, 171–72
Cline, Lisa, 181
Clinton, Bill, 69
Clinton Foundation, 52–53
CloudPets, 9
coal industry, 170–71
Coca-Cola Company, 167, 169
CoComelon (YouTube show), 19, 212
"collectibles," 83–88
  LOL Surprise dolls, 85–86, 87
  plastics packaging, 85–86
  plastics pollution, 83–88
  Pokémon and, 30, 84–85, 105–6
  See also "unboxing videos"
Color of Change, 231
Colorado State Board of Education, 234
Colorado State University, 167–68
colorism, 153
commercialism, definition of, 60–61
Common Sense Media (CSM), 94, 216
communicative robots (or "sociable bots"), 141–43, 231
consumer capitalism, 81
consumer protection movement, 166
Consumers Union of the United States, 167
Consuming Kids: The Hostile Takeover of Childhood (Linn), 2, 4, 7, 47, 60, 91, 120, 160
Cook, Tim, 68
Cookie Monster, 27
"cooperative sandbox games," 49–50, 53–55, 218–19
COPPA. *See* Children's Online Privacy Protection Act (COPPA)
corporate capitalism, 81
The Corporation (film), 114–15
Corporation for Public Broadcasting (CPB), 52–53
COVID-19 pandemic, xiii–xv, 55
  and children's screen time, 39–40
  and the digital divide, 40–41
  and Facebook, xiv
  students' use of edtech products, 189, 195–96
  and systemic racism, xiii–xiv
  and Trump administration, xiv–xv
  video chats and puppet therapy during, xiii, 33, 138, 194
credit card companies, 172–75, 180
critical thinking, xiv, 21, 69, 168, 180, 240
CTA Digital's iPotty, 107

data mining, 47–48, 81
DDB Worldwide, 69
"deep culture," 7
default settings for online and digital services, 77–78, 200–201, 204
  autoplay settings, 77–78, 128, 204
Defending the Early Years (DEY), 209, 256
Delphi, 231
democratic citizenship, public education and, 168–69, 172
deregulation of children's television, 3, 29
developmental psychology, 12–13
  brains and early experience, 12–13, 81–82, 210
  as tool of digital marketing and advertising, 51–52, 112
Dewey, John, 168–69
digital beauty filters, 5, 73, 153
digital devices, 4–5, 125–43
  children's leisure time with, 14, 40, 100, 104, 129
  communicative robots (or "sociable bots"), 141–43, 231
  disruption to families, 130–33
  effects on conversations between parents and children, 130–31, 132–33
  lure of convenience, 222–23
  parasocial relationships and children's emotional attachments to, 142–44
  self-control and, 129–30
  sleep and, 104, 224
  touch screens, 40, 215
  voice-activated personal assistants, 133–41, 151
  See also edtech industry; screen time, children's; smart devices; smartphones
digital divide, 40–41
digital "learning games." *See* edtech industry; Prodigy (freemium math app)
digital personal assistants, 133–41, 151
  Amazon's Alexa, 133–34, 137–41, 151
  Echo Dot Kids Edition, 134–41, 151
digital "sandbox" games, 49–50, 53–55, 218–19. *See also* Fortnite; Minecraft

digital toolbox and Digital Tools Criteria Checklist, 245–48
Digital Wellbeing Bill (Minnesota), 236–37
Digital Wellness Institute, 256
Discover, 172–73, 177
Disney, 3, 141–42
   Baby Einstein videos, 4, 199
   films and racially stereotyped images, 159–60, 161
   teaching materials for schools, 164–65
   toys and media-linked characters, 28
Disney+, 159
Disney Junior, 28
Disney Remix, 139
Dora the Explorer, 142–43
Dory (Disney character), 28
DreamWorks, 139, 141
drug advertisements on television, 92
"dumb" watches, 220
Dumbo (film), 159

e-books, 132–33, 214
   enhanced, 132
   See also e-readers
e-cigarettes, 165
"e-people" (e-girls and e-boys), 62–63
e-readers, 127–28, 134, 214, 221
eBay Inc., 148
Echo Dot, 134, 141. *See also* Amazon Alexa (voice-activated digital assistant); Echo Dot Kids Edition
Echo Dot Kids Edition, 134–41, 151
   and Amazon Kid+, 136
   Amazon Parent Dashboard, 136
   data collection and privacy concerns, 135–36
   drop-in feature, 136
   4th Gen hardware, 141
   personalized recommendations, 137–40
edtech industry, 71, 181–96, 231–33
   and COVID-19 school closings, 189, 195–96
   edtech policy at Acton-Boxborough Regional School District, 232–33, 245–48
   "free" apps ("freemiums"), 46, 119, 181, 186–87, 217
   gamified products, 192–94
   hardware dumping and brand loyalty, 183–84
   inBloom administrative platform and data collection, 233–34

"personalized learning," 184–85, 190–94
Prodigy app, 181, 184–89, 192–94, 232–33
research on effectiveness of, 184–85, 187–88, 190, 191–92, 194–95
teachers as brand promoters, 71
three categories, 183–85
Education Week, 177
Einstein, Albert, 22
Elle (magazine), 90
Elmo, 13, 212, 217
Entertainment Junction, 78
European Union, 38, 115
exercise, physical, 103

Facebook, 5, 63, 73, 74–76, 145, 148, 152
   and COVID-19 pandemic, xiv
   fake news, xiv, 9
   Messenger Kids (MK), 74–75, 219
   racism and racial biases, 152–53
   revenue, 130
   targeted advertising to teens, 74–76
   See also Meta
Facing the Screen Dilemma (Linn, Almon, and Levin), 13
Fairplay, 3–4, 199–200, 202, 228, 231, 239
   campaign to convince Scholastic to stop distributing SEMS, 170
   campaign to prevent Mattel from launching Aristotle, 36, 38
   campaign to stop false advertising at Your Baby Can Read, 145–46
   complaint to FTC about Prodigy, 187, 233
   Screen Time Action Network, 234, 253
   statement on edtech and "personalized learning tools," 185
   website, 253
   Worst Toy of the Year Award, 107
"Faith Leaders Letter to Mark Zuckerberg" (2022), 239
Family Online Safety Institute (FOSI), 179
Family Room (market research firm), 120–21
Fantasy Rush (fantasy football game), 4
Federal Communications Commission (FCC), 29
Federal Trade Commission (FTC), 29, 187, 199, 202–4
   and Amazon's Echo Dot Kids Edition, 134–35

Federal Trade Commission (FTC) (*cont.*)
  enforcement of COPPA, 135, 202–3
  Fairplay's complaint about Prodigy, 187, 233
  and false advertising at Your Baby Can Read, 145–46
  Google and YouTube investigations, 4, 177
  and legislation to end Big Tech's exploitation of children, 135, 202–4
  and television advertising to children, 113
Filene, Edward, 168–69
Finance Parks, 163–64
financial literacy materials produced by credit card companies, 172–75, 180
Finding Dory (film), 28, 164
Finding Nemo (film), 28
Fisher, Mike, 184
fitness trackers, 103
5Rights Foundation (UK), 231, 254
Flint, Michigan, 65
Florida Atlantic University, 69–70
Floyd, George, xiii–xiv, 150, 152
food industry
  Big Food, 198
  Impossible Foods and in-school teaching materials, 175–76
  in-school advertising, 165, 167, 172, 175–76, 180
  junk-food marketing to kids, 26, 82, 198
Foolproof Foundation, 257
Forbes, 67, 85
Fortnite, 49–50, 54, 110, 118, 185, 186, 218–19
fossil fuel industry
  corporate-sponsored educational materials (SEMS), 170–72, 179–80
  greenwashing, 190
Franz, Rachel, 207
"free" apps ("freemiums"), 46, 119, 181, 186–87, 217
  and nagware, 46, 186–87, 189
  Prodigy, 181, 184–89, 232–33
Freed, Richard, 52
frictionless commerce, 77–78, 81
Friends of the Earth, 170
Frozen Sing!, 141
Fujioka, Robb, 37
"fun" and gamified edtech products, 192–94
Futuresource Consulting, 184

gambling, 43, 60, 86, 95
gamification
  edtech products and making learning "fun," 192–94
  Pokémon Sleep app, 103–5
  Pokémon Smile and tooth-brushing, 105–7
  See also Prodigy (freemium math app); rewards
Gates, Bill, 218
Geico, 163
General Electric, 182
General Motors, 166
Generation Z, 61–62, 64–65
Geoffrey the Giraffe (Toys "R" Us mascot), 57
Georgetown University's Institute for Public Representation, 145–46, 202
girls, teenage
  digital beauty filters and body image, 73, 153
  "e-girls," 62–63
  Google's search engines and racist search results, 149–51
  and Instagram, 74, 75
  and social media, 73–75
  and teen tribes, 62–63
Godin, Seth, xiii, xv
Golin, Josh, 3, 36, 40, 189, 223
"good enough" mothers/families, 14, 30–31, 37
Google, 145–51, 165, 176–79, 181, 202
  advertising revenue, 148
  digital assistant, 134
  FTC investigations, 4, 177
  hardware dumping in schools, 184
  sponsored educational materials (*Be Internet Awesome*), 176–79
  tracking and data collection, 4, 177–78, 183
  Workspace for Education, 183
  See also Google Search; YouTube
Google Apps for Education, 177–78
Google Maps, 177
Google Play Store, 119, 216
Google Search, 145–51
  algorithms, 146–51
  pornography and, 149
  racism, 146–51
Grande, Ariana, 90–91
Great Recession (2008), 65, 121
Greater Good Science Center, 23

INDEX

greenwashing, 190
Gulf of Mexico oil spill, 171
Guynn, Jessica, 150

Hampton, Maree, 235–37
happiness and consumption, 7–8, 79, 88, 98, 168
Happy Meals and Happy Meal toys, 84
Harding, Vincent, 230
hardware dumping, 183–84
Harlow, Harry, 36–37
Harrell, D. Fox, 145
Harris, Tristan, 129–30
Harry Potter, 139n, 185
Harvey, Peggy, 245n
Hasbro, 4, 57–58, 86, 199
Haworth, Elliott, 111
Heartland Institute, 171
Heinz, 166
Hello Barbie, 142
Hello Kitty, 28
Hershey, 166
Heschel, Abraham Joshua, 22–23
Heyer, Ken, 131
High Fidelity (film), 89
Hirsh-Pasek, Kathy, 11
Hispanic Barbie, 155–57, 162
Historical Racialized Toys in the United States (Barton and Somerville), 158
"holding" environments, 15–16, 19–20
holiday gift-giving, 226
host-selling, 140
The House at Pooh Corner (Milne), 138
Hughes, Lucy, 114–15
Human Subjects Review, 114

Idell, Cheryl, 113
iGen (Twenge), 43, 235
Iger, Bob, 57
Ignite Technologies, 133–34
"impinging" environments, 16–17
Impossible Foods, 175–76
inBloom, 233–34
India, 117, 157
infinite scroll, 46, 73, 208
influencers, social media, 5, 72, 86–88, 204
Ingersoll Lecture at Harvard Divinity School, 197
Initiative at Media, 114–15
Instagram, 5, 63, 72–76, 204, 219
  algorithms and racial biases, 151–53
  influencers, 72, 87–88

Instagram for Kids, 74–75, 219, 239
  photo editing feature, 73, 153
Instagram for Kids, 74–75, 219, 239
Instituto Alana (Brazil), 231
intermittent reward schedules, 46–47, 185
International Society for Technology in Education (ISTE) Standards for Students, 178–79
Internet of Things (IoT), 45, 47, 133. *See also* smart devices
intrinsic motivation, 8, 97–103, 107–8
iPads, 35, 107, 129, 235
  distribution in schools, 163–64, 183–84, 235
  Finance Parks and in-school advertising, 163–64
  YouTube videos of babies playing with, 17, 19
iPotty, 107
Iran, 116–17
Ishihara, Tsunekazu, 104
iTunes, 93

Jain, Amrita, 157
Jefferson County, Colorado, 233–34
Jobs, Steve, 129
Johns Hopkins University, 188
Johnson, Lyndon, 69
Julie Albright (American Girl), 139
The Jungle Book (film), 164
Junior Achievement Finance Parks, 163–64
Juul, 165

Kahn, Lina, 203
Kaji, Ryan, 72
Kasser, Tim, 91
Keltner, Dacher, 23
KIDS Act (Kids Internet Design and Safety Act), 204
Kids Online Safety Act (KOSA), 204
Kids PRIVCY Act, 203–4
"kids tech industry," 4
Kindle e-reader, 134
Koch, Charles, 171, 172
Koch Industries, 171–72
Kohn, Alfie, 93, 191
Ku Klux Klan, 158

Lady and the Tramp (film), 159
Lancet (journal), 83
language acquisition, children's, 41–42
Lanier, Jaron, 130

Larian, Isaac, 85, 86
Latinx communities, 155–57
legislative efforts to limit Big Tech and
    protect the rights of children,
    199–205, 230
  Britain's Age-Appropriate Design Code,
    200–201, 203
  COPPA, 135, 201–3
  KIDS Act, 204
  Kids PRIVCY Act, 203–4
  KOSA, 204
  Minnesota's Digital Wellbeing Bill,
    236–37
  state legislatures, 204–5, 236–37
Lego
  apps, 4, 93–94, 95, 100, 119
  blocks, 93, 100
  Ninjago empire, 94
  revenue, 58
Lego City apps, 93–94, 95, 100
Lego Ninjago: Masters of Spinjitzu
    (TV series), 94
Lego Ninjago Movie, 94
Lego Ninjago Movie Video Game, 94
Lesley University (Massachusetts), 209
Let Grow, 255
Levin, Diane, 13
"likes," 46, 47, 63, 72, 73, 152, 153
Little, Brown and Company, 139
LiveMore, ScreenLess, 235–37, 257
Log Off, 237, 257
LOL Surprise dolls, 25n, 85–86, 87
London Times, 62
Look up, 257
loot boxes ("blind" boxes or bags), 86, 95
Lost Kitties, 86–87
Luddites, 34–35

Ma, Yo-Yo, 25
Macy's, 58
Marcum, J. Paul, 131
"market segmentation," 51
Markey, Ed, 38
Massachusetts Institute of Technology
    (MIT), 44
Masters of Spinjitzu online games, 94
"material culture," 154, 158
materialism and materialistic values, 8, 79,
    82–92, 96–100
  advertising and exacerbation of, 83
  Ariana Grande's "7 Rings," 90–91
  buying happiness, 7–8, 79, 88, 98, 168

and Channel One News in schools,
    182–83
  and children's relationships, 89
  climate change and consumption, 83, 88
  and "collectibles," 83–88
  defining, 83
  external rewards, 8, 97–102, 104–8,
    193, 194
  retail therapy, 90–91
Mattel, 57–58, 142
  American Girl Dolls, 139
  Aristotle smart device, 36–38
  Barbie, 85, 142, 155–57, 162
McDonald's, 177, 199
  as "American institution," 59, 60
  brand loyalty, 66–67
  fast food SEMS and advertising in
    schools, 165, 167, 172
  Happy Meal toys, 84
  "McTeachers' Nights," 165
  school-based nutrition program, 172
  social media followers, 63
McNeal, James, 66
Me Too movement, 152
Mead, Margaret, 230
media literacy, 205, 222
Messenger Kids (MK), 74–75, 219
Meta, xiv, 46, 74–76, 151–52, 176, 240
  algorithms and racial biases, 151–52
  Russian interference in 2016 election,
    176
  See also Facebook; Instagram;
    Instagram for Kids
MGA Entertainment, 25n, 85–86
Microsoft, 53–54
millennial parents and families, 120–23
Minecraft, 53–55, 110, 218–19
minimalism, 88
mining (data mining), 47–48, 81
Minnesota State Legislature, 204–5, 236
Mister Rogers' Neighborhood, 3
MIT Technology Review, 190
Molnar, Alex, 167–68, 197
mommy bloggers, 87–88
Monster High fashion dolls, 59
Montgomery Ward catalog, 158–59
Morrison, Toni, 155
Mosseri, Adam, 75
motivation
  external rewards, 8, 97–102, 104–8,
    193, 194
  intrinsic, 8, 97–103, 107–8

MTV, 114
Museum of Fine Arts (Boston), 120
Museum of Science (Boston), 119–20
My Little Pony (TV series), 19, 28
Myers, KK, 235–37

Nabisco, 105–6
nagging, 78, 109–24, 227
    and child-targeted advertising, 78, 109–24
    and early adolescence, 110
    and "free" apps, 46, 119, 217
    harm to families, 78, 82, 109–11, 115, 117–20, 122–24, 231
    international market research on strategies to encourage, 116–17
    marketers' work to neutralize parents as gatekeepers, 111–12, 122–23
    and millennial parents, 120–23
    "The Nag Factor Study" (1998), 113–15
    nagware and pop-up ads, 46, 186–87, 189
    "persistence nagging" and "importance nagging," 113
    "pester power" ("nag factor"), 112–15, 117
    saying "no" to children, 227
    and Toys "R" Us free camps, 60
nagware (pop-up ads), 46, 186–87, 189
A Nation at Risk (1983 report), 166
National Association of Manufacturers, 182
National Education Association, 166
National Education Policy Center at the University of Colorado, 192, 195
National Football League (NFL), 4
National Parent Teacher Association (PTA), 179, 238
national parks, advertising in, 7
National Public Radio, 68
nature experiences
    advertising in national parks, 7
    awe and, 23–24
    spending tech-free time outdoors, 224
Nestle, Marion, 163
neurotypical children, 9, 12
New York Times, 36, 38, 54, 68, 71, 105, 119, 131, 170, 184
Newton, Isaac, 22
Niantic, 102–3
Nickelodeon, 86–87, 90, 114, 142–43
Nike, xiv, 121, 169
1984 (Orwell), 68

Nineteenth Amendment to the U.S. Constitution, 230
Nintendo, 84, 102, 116, 167. *See also* Pokémon
Noble, Safiya Umoja, xv, 148–49, 151
nonprofits, corporate funding of, 222
notifications. *See* push notifications

Obama, Barack, 69
Oklahoma Energy Resources Board (OERB), 171
Oklahoma schools, 171
Oreos, 105–6
Organization for Economic Co-Operation and Development, 184
Orwell, George, 68
Oxford English Dictionary, 91, 105, 194

Pakistan, 117
Paramount Global, 143
Paramount Pictures, 114
parasocial relationships, 142–44
parents and caregivers, suggestions for. *See* resistance parenting suggestions
Parents Together, 258
Paw Patrol (YouTube show), 119, 212
pediatric public health organizations, 209–10
Peppa Pig (British animated series), 48, 212–13
Pepsico, xiv, 167, 169
"personalized learning"
    and "constructing knowledge," 191–92
    edtech and, 184–85, 190–94
    Fairplay statement on, 185
    Kohn on teachers and true, 191
    and "making meaning," 191–92
    and progressive education, 191–92
personalized recommendations, 45, 48–49, 50, 137–41. *See also* algorithms
persuasive design techniques, 5, 9, 45–46, 126, 132, 185
Peter Pan (film), 159
Pew Charitable Trusts, 61, 120
Pineapple Lounge, 111
Pirates of the Caribbean (film), 164
Pizza Hut, 165
plastics
    and climate change, 85, 87–88
    collectible toys, 83–88

ocean pollution, 85
packaging, 85–86
play, creative, 14–17, 25–29, 42, 87, 100, 193, 210–11
   declines in, 29
   online sandbox games and monetization of, 49–50, 53–55, 218–19
   play-based preschools, 192
   value of playing with friends in person, 218–19
   Winnicott on best environments for, 14–17
PlayCon 2018 (Toy Industry conference), 57–58, 61–62
playdates, tech-free, 225
Plessy v. Ferguson (1882), 158
Pokémon, 30, 84–85, 102–6, 185
   branding and, 30
   collectibles craze, 30, 84–85, 105–6
   theme song chorus ("gotta catch 'em all"), 30, 85
Pokémon GO, 102–4
Pokémon Oreos, 105–6
Pokémon Sleep, 103–5
Pokémon Smile, 105–6
Political Tribes (Chua), 61–62
pop-up ads (nagware), 46, 186–87, 189
poverty
   children and, 65
   COVID-19 disparities, xv
   materialistic values and, 97
   millennial families and, 121
precautionary principle, 38–39, 209
presidential branding, 67, 69–71
privacy
   Apple, 68
   Children's Online Privacy Protection Act (COPPA), 135, 201–3
   data mining, 47–48, 81
   Echo Dot Kids Edition, 135–36
   edtech and inBloom administrative platform, 233–34
   Google tracking and data collection, 4, 177–78, 183
   online behavior tracking, 47–49, 137, 177–78, 183
   smart devices and surveillance/data collection, 36, 47, 133–36, 177–78
   YouTube, 4, 47–48
Prodigy Education, 185–86
Prodigy English, 186

Prodigy (freemium math app), 181, 184–89, 192–94, 232–33
   nagware and pressures to upgrade, 181, 186–87, 189
   question of effectiveness at teaching math, 186, 187–89, 232
   as social network, 189
professional organizations, corporate funding and, 238
progressive education
   Dewey's philosophy of, 168–69
   and personalized learning, 191–92
Psychological Science in the Public Interest (journal), 194–95
Psychology Today, 191
Public Broadcasting Service (PBS), 52–53
public education
   American democratic citizenship, 168–69, 172
   Dewey's philosophy of, 168–69
   purpose of, 168–69, 172
   See also edtech industry; schools, corporate advertising in
public media (commercial-free), 52–53, 216
Publicis Groupe, 117
Punished by Rewards (Kohn), 191
puppet therapy, 3, 27
   characters' behavior as models for children, 20
   and parasocial relationships, 143
   video chats during COVID-19 pandemic, xiii, 33, 138, 194
push notifications, 46, 200, 204, 220–21
Pussy Cat Dolls, 4
Putnam, Robert, 132

racism, racial biases, and racial stereotypes, xv–xvi, 146–62
   Alexa's search feature, 151
   algorithms and, xv, 146–53
   children's stories and racialized dolls and toys, 153–62
   and COVID-19 pandemic, xiii–xiv
   and digital beauty filters, 153
   Disney films, 159–60, 161
   Google's search engines, 146–51
   name searches and search terms (e.g., "Black girls"), 146–47, 149–51
   and postslavery history, 158–59
   and social media sites, 151–53
   white dolls and, 153–57

Radesky, Jenny, 119, 125
radio advertising (mid-twentieth century), 132
Raffi Foundation for Child Honoring, 258
Ranking Digital Rights, 258
reading books
  differences between using smartphones and, 126, 127–28, 132–33
  e-books and e-readers, 132–33, 134, 214, 221
  reading to children, 132–33, 214
Reagan, Ronald, 29, 69, 166
Reclaiming Conversation: The Power of Talk in a Digital Age (Turkle), 133
Reinhard, Keith, 69
religious organizations and spiritual leaders, 71, 238–39
Reset Australia, 75
resistance parenting suggestions, 207–28
  age-appropriate and commercial-free apps, 216–17
  be wary of convenience, 222–23
  family plan/agreement about tech use at home, 222
  helping kids resist marketing-induced behaviors and values, 225–27
  infants and toddlers, 212–14
  middle childhood/elementary grades, 217–19
  preschool and kindergarten, 214–17
  reading to kids, 214
  reducing your own time with tech, 220–21
  screen time limits, 43, 53, 209–11, 213, 215, 223
  and six principles of child development, 209, 210–11
  smartphones, 218, 220–21, 225
  social media sites, 219
  tech-free, commercial-free family time, 223–24
  tech-free solutions to boredom, 224–25
  tech-free time outdoors, 224
  tech-free zones in the home, 224
  things to consider in making decisions for your family about tech and commercial culture, 222–25
  toys, 213–14
retail therapy, 90–91
Rethinking Schools, 170

rewards
  external (material), 8, 97–102, 104–8, 193, 194
  intermittent (persuasive design technique), 46–47, 185
  and intrinsic motivation, 8, 97–103, 107–8
Rideout, Vicky, 1
Roblox, 218
Rodgers and Hammerstein, 90
Rogers, Fred, *3*, 25
Roof, Dylann, 149
Rowling, J.K., 139n
Royal College of Paediatrics and Child Health, 211
Rubenstein, Rheta, 186
Ryan's World (Ryan's ToysReview), 72, 204

Salesforce IoT, 45
same-sex marriage legalization, 230
Sanrio, 28
Scholastic, 170
schools, corporate advertising in, 163–80
  Channel One News, 182–86
  corporate-sponsored "educational" films (1950s), 182
  corporate teaching materials (SEMS), 164–66, 169–80, 199
  credit card companies and financial literacy materials, 172–75, 180
  Disney, 164–65
  fast food industry, 165, 167, 172, 180
  financially-strapped schools, 166–67, 300–1n20
  fossil fuel industry, 170–72, 179–80
  Google's *Be Internet Awesome*, 176–79
  Impossible Foods, 175–76
  instilling brand loyalty, 164, 165, 172, 184
  local level efforts to change, 231–34
  "partnerships," 164, 167
  science materials that deny climate change, 171–72
  and true purpose of public education, 168–69, 172
  See also edtech industry
Schwartz, Shalom, 96
Scooby-Doo, 67
Scotland, 118
Screen Free Week, 234, 254

screen time, children's, 39–43, 52–53,
    118–19, 209–11, 213
  babies' and toddlers' interaction with
    screens, 17–19, 39, 41–43, 213
  and children of tech executives, 129, 218
  and COVID-19 pandemic, 39–40
  and language acquisition, 41–42
  online videos, 40
  recommendations, 43, 53, 209–11, 213,
    215, 223
  state bills to limit in early childhood
    settings, 204–5
  touch screens, 40, 215
  TV watching, 40, 42, 100
Screen Time Action Network at Fairplay,
    234, 253
search engine optimization (SEO)
    companies, 148
search engines
  Alexa's search feature, 151
  algorithms and, 146–53
  Google Search, 145–51
  name searches and search terms,
    146–47, 149–51
  racial and ethnic biases, 146–53
  and SEO companies, 148
Searle, Mike, 61
Sears catalog, 158–59
self-regulation, 21, 42, 193
SEMS ("Sponsored Educational
    Materials"), 164–66, 169–80, 199
  financial literacy materials produced by
    credit card companies, 172–75, 180
  and fossil fuel industry, 170–72, 179–80
  Google's *Be Internet Awesome*, 176–79
  Impossible Foods and climate change
    education, 175–76
  science materials that deny climate
    change, 171–72
  See also schools, corporate advertising in
Seneca Falls Women's Rights Convention
    (1848), 230
The Sense of Wonder (Carson), 23
September 11, 2001 terrorist attacks, 91
Sesame Street, 27, 131
Sesame Workshop, 52–53, 131
"7 Rings" (song), 90
sexualized toys, media, and clothing, 4,
    26, 59, 85, 156
Share Save Spend, 258
Shell Oil Company, 165, 171, 182
Shenoy, Neal, 136

Shine, Nora, 232–33, 234
Shrek, 67, 142
Silent Spring (Carson), 23
Siri, 134
ŠKODA Auto, 117
slavery, 158–59, 230
sleep, 103–5, 224
  digital devices and, 104, 224
  Pokémon Sleep app, 103–5
slogans, advertising, xiv
smart devices, 36–38, 45, 47, 133–41
  Amazon's Alexa, 133–34, 137–41, 151
  Aristotle, 36–38
  attachment and, 36–38, 141–43
  digital personal assistants, 133–41, 151
  and "dumb" watches, 220
  Echo Dot Kids Edition, 134–41, 151
  Internet of Things (IoT), 45, 47, 133
  privacy and data collection, 36, 47,
    133–36, 177–78
  See also smartphones
smartphones, 35, 45, 66, 100, 105,
    126–28, 130–33
  differences between reading books and
    using, 126, 127–28, 132–33
  effect on conversations between parents
    and children, 130–31, 132–33
  and internet access, 40–41
  marketed to children, 130–31
  parents' and caregivers' overuse of, 41,
    125–27, 221
  resistance parenting advice, 218,
    220–21, 225
Snapchat, 5, 73, 152–53, 202, 204
social change, steps toward systemic,
    198–99, 208, 229–40
  at the local level, 231–35
  Minnesota's Digital Wellbeing Bill,
    236–37
  organizations working to change laws
    and policies, 231, 234–35
  organizing screen-free weeks in
    communities, 234–35
  religious organizations and spiritual
    leaders, 238–39
  schools and school districts, 231–34
  young people's advocacy groups,
    237–38
social media sites, 5–6, 40, 72–76,
    151–53, 219
  and adolescents' mental health, 73–75,
    219

algorithms and racial biases, 151–53
brand tribes and marketing, 63–65
girls and, 73–75
influencers, 5, 72, 86–88, 204
intermittent rewards, 46–47
"likes," 46, 47, 63, 72, 73, 152, 153
resistance parenting recommendations, 219
self-branding, 5, 72, 76, 219
Somerville, Kyle, 158
The Sound of Music (film), 90
South Africa, 117
Spears, Britney, 3
Spiderman, 13, 67, 121
SpongeBob SquarePants, 138, 142
SpongeBob SquarePants Challenge, 139, 140
Sponsored Educational Materials. *See* SEMS ("Sponsored Educational Materials")
Sprinkle, David, 109
Sprite, xiv
St. Jude Children's Research Hospital, 106
Stanford University, 44
Star Trek: The Motion Picture, 84
Star Wars, 141
STEM (Science, Technology, Engineering, and Mathematics) education, 170–71
Stonewall Uprising (1969), 230
Story of Stuff Project, 38
Strickland, Rachael, 233–34
Student Data Privacy Project, 181, 258
surveillance and online behavior tracking, 47–49, 137, 177–78, 183
surveillance capitalism, 80–81

Target, 19, 59
"target marketing," 51
tech jargon in lay language, 45–46
"teen tribes," 62
television
 ban on children's advertising, 113, 140
 Channel One News, 182–83
 children's TV watching, 40, 42, 100
 deregulation of children's, 3, 29
 host-selling and children's programming, 140
 public (commercial-free), 52–53, 216
 racist images and stereotypes, 153, 159–60
 story-based programs and films, 216

There Is a River: The Black Struggle for Freedom (Harding), 230
Tijuana, Mexico, 153–54, 156–57
TikTok, 72–73, 152, 202, 219, 238
Today show, 145–46
toilet training, 107
Tony the Tiger (Kellogg's mascot), 57
tooth-brushing (gamification and rewards), 105–7
touch screens, 40, 215
toy companies, 29–30, 57–60
 Hasbro, 4, 57–58, 86, 199
 and influencers, 5–6
 Lego, 58
 Mattel, 36–38, 57–58, 139, 142
 and plastic "collectibles," 83–88
 PlayCon 2018 (industry conference), 57–58, 61–62
 and Toys "R" Us bankruptcy, 57–61
 "unboxing videos," 5–6, 85, 177, 204, 227
toys
 brand-licensed, 13–14, 29–30
 and children's creative play, 25–29, 59, 87
 electronic, 14, 19, 24
 "interactive," 17, 19
 media linked characters and, 27–28, 29–30, 212–13
 plastic collectibles, 83–88
 racialized and racist, 153–62
 recommendations for healthy development of babies and toddlers, 213–14
 Toys "R" Us marketing and, 57–61
 violent or sexualized, 4, 26, 59, 85, 156
Toys "R" Us, 57–61
Trend Hunter, 61
tribes, 61–65
 brand tribes and social media, 63–65
 "e-people," 62–63
 Gen Z and, 61–62, 64–65
 "teen tribes," 62
Trolls Hair Huggers, 86–87
Truce, 259
Trump, Donald, xiv, xv, 67, 70–71
Tulsa Race Massacre (1921), 158
Turkle, Sherry, 54–55, 133, 141
Turning Life On, 235, 259
Twenge, Jean M., 43, 235
Twisty Petz, 86–87
Twitchell, James, 60–61

Twitter, 145, 153
Ty Corporation, 84

"unboxing videos," 5–6, 85, 86–87, 177, 204, 227
UNICEF, 91
Unilever, 64
United Nations Convention on the Rights of the Child, 112, 199
United Nations Education, Science and Culture Organization (UNESCO), 38–39
University of Michigan, 119, 125, 154
University of Michigan–Dearborn, 186
University of Missouri, 182–83
The Unlucky Adventures of Classroom Thirteen (book series), 139
USA Today, 150
user experience (UX design), 45. *See also* persuasive design techniques

values, learning, 95–100
   materialistic, 96–100
   resistance parenting (helping kids resist destructive marketing-induced values), 225–27
   universal system of basic human values, 96–97
   See also rewards
Vampirina (Disney Junior character), 28
Verizon, 131
Viacom, 114, 141–42
Victorious (Nickelodeon series), 90
violence in children's media and games, 26, 39, 48, 59, 82, 204
Visa, 172–75, 177
Vivid Brand, 63
VJR Consulting, 1

Volkswagen, 163
VTech, 9

Wait Until 8th, 218, 259
Wall Street Journal, 58, 74, 103, 148, 177
Walmart, 19, 59
Warner Brothers, 138
Washington Post, 60, 113
Wells Fargo, 163
Western Media International, 113–15
Westerveld, Jay, 190
white supremacists, 149, 152, 158
Why Scientists Disagree About Global Warming (Heartland Institute book), 171
Wii, 116
Winnicott, D. W., 14–17, 30
Wired, 72
Wizarding World Book Quiz, 139
wonder. *See* awe and wonder
World Economic Forum, 85
Worrollo, Emma, 111–12

Your Baby Can Read, 145–46
YouTube, 5–6
   default autoplay setting, 77–78
   FCC investigations, 4, 177
   influencers, 5–6
   predictive algorithms and "recommended" videos, 48
   racism and racial biases, 152–53
   Ryan's World, 72, 204
   self-posted videos, 6
   "unboxing videos," 5–6, 177
YouTube Kids, 4, 77n, 121, 177

Zavattaro, Staci, 69–70
Zinn, Howard, 179
Zuboff, Shoshana, 80–81
Zuckerberg, Mark, 74, 239, 240

About the Author

Susan Linn is a psychologist, award-winning ventriloquist, and a world-renowned expert on creative play and the impact of media and commercial marketing on children. She was the Founding Director of Campaign for a Commercial-Free Childhood (now called Fairplay) and is currently a lecturer on psychiatry at Harvard Medical School. The author of *Consuming Kids* and *The Case for Make Believe* (both published by The New Press), she lives in Brookline, Massachusetts.